YANKEE MUSLIM

The Asian Travels of Mohammed Alexander Russell Webb

Edited by

Brent D. Singleton

The Borgo Press
An Imprint of Wildside Press

MMVII

**Stokvis Studies in Historical
Chronology and Thought
ISSN 0270-5338**

Number Seven

FIRST EDITION

CONTENTS

YANKEE MUSLIM, ED. BY BRENT D. SINGLETON

PREFACE

The 1892 travel journals of Mohammed Alexander Russell Webb provide the reader with a glimpse into the mind of an interesting, yet forgotten, character of late nineteenth century America. The travel journals are the first Webb ever wrote, and the opening of the initial volume is somewhat rough as he learns the art of keeping a journal. The writing quickly gains momentum when he begins to describe Singapore. The succeeding three months of journal entries provide useful descriptions and analysis of his travels and personal encounters. The end result is an important document for the study of early American Muslim history.

The physical journals were purchased by the Duke University Rare Book, Manuscript, and Special Collections Library in 1980, where they currently reside under the title of the Alexander Russell Webb Papers. The journals consist of two volumes, each 144 pages in length and written in both pencil and ink, which correspond more or less to Webb's traveling (pencil) and his stay in cities (ink). The first volume covers August 29—October 19, 1892, and describes Webb's leaving Manila and traveling to Singapore, Penang, Rangoon, and the beginning of his stay in Calcutta. The second volume covers the dates October 20—December 15, 1892, finishing his time in Calcutta as well as travel to Patna, Benares, Bombay, Poona, Hyderabad, Madras, and Agra. Webb remained in India an additional two weeks, but the whereabouts of subsequent travel journals for this period and after is unknown.

Three lectures that Webb delivered in Bombay, Hyderabad, and Madras have also been included in this work.

They have been reproduced from Webb's book, *The Three Lectures*. Madras: Lawrence Asylum Press, 1892.

Editorial Methods

The editor has endeavored to reproduce precisely what Mohammed Alexander Russell Webb wrote in his travel journals. Throughout the work Webb often used British spelling conventions, phonetic spellings of names and foreign words, and he also simply misspelled words, these have not been corrected. Nor have his inconsistent forms of punctuation and capitalization. All hyphenated words that continue on a second line were combined where appropriate and instances of double word entries have been deleted. Webb used a small 'x' for a period and a dot for a comma, these have been normalized with proper periods and commas. As well, he often employed dots after numbers, these have been removed.

Notes have been provided where the editor felt modern readers or those unfamiliar with the subject matter might benefit. Most foreign and uncommon words appear in the glossary, however, in a few instances the *Oxford English Dictionary* or other common reference works are quoted in a note. Basic biographical information has been supplied for those individuals that could be identified in other sources. For those luminaries with extensive biographies, only the highlights of their lives and relation to Webb are mentioned. The biographical notes for persons mentioned in both the introduction and the journals are only listed at the first mention of their name in the journal. Unfortunately, there are many figures that could not be identified beyond Webb's account and therefore no note is provided in such cases.

Acknowledgments

This book could not have been completed without the assistance of many people and institutions. First, I would like to thank Michael Burgess for his support and assistance in bringing this project to publication. I would also like to thank my colleagues at the California State University, San Bernardino, John M. Pfau Library for their support, especially Bonnie Petry and Stacy Magedanz for their proofreading assistance. Special thanks to Lee Bayer, Annemarie Hopkins, and the Interlibrary Loan staff at CSUSB for their work in tracking down resources. As well, the CSUSB Professional Awards Committee for funding part of my research.

Thanks to the following individuals for supplying materials: Dr. Zahid Aziz, Editor, "The Light"; Nina Schneider, New York Public Library; Jackie Bunker-Lohrenz, Meadowlands Museum; Laura Gerwitz, Mamiya Medical Heritage Center; David Wietersen, Theosophical Society Pasadena; Dr. Cezmi Eraslan, Istanbul University; Dr. Milton O. Gustafson, National Archives and Records Administration; Jason Stratman, Missouri Historical Society; Dr. Susan Holly, U.S. State Department; Dr. Hamid Rahman; Dr. Omar Khalidi; Dr. Ken Robbins; and Dr. Dany Doueiri.

The following Institutions were kind enough to supply materials for this project: Rare Book, Manuscript, and Special Collections Library at Duke University; Library of Congress; University of Rochester; State Historical Society of Missouri; St. Louis Public Library; Western Reserve Historical Society; The British Library; University of California, Los Angeles; and the University of Washington.

Finally, I would like to thank my family for their support and understanding, Carmon, Imani, and Anisa.

This book is dedicated to

Carmon, Imani, and Anisa

INTRODUCTION

MOHAMMED ALEXANDER RUSSELL WEBB

On February 16, 1893 the White Star Line's steam-ship *Majestic* moored in New York harbor fresh from a six-day voyage from Liverpool, England. Among the throngs of European and American travelers was forty-six-year-old Mohammed Alexander Russell Webb, ex-consul to the Philippines, former newspaperman, and convert to Islam. He had returned to the United States, after touring India and spending more than five years abroad, to ignite a top-down Islamic revolution among the intelligentsia of Gilded Age America. He had taken up the challenge to rid the county of its prejudiced view of Islam and gain a foothold for the faith within the nation's spiritual consciousness. Webb and his East Indian financial backers envisioned a full-fledged Islamic propaganda movement, mosques in every major city, and the eventual establishment of Muslim colonies. He used his knowledge of the newspaper business to secure coverage for himself and his cause as he stormed across the country lecturing and publishing. Dubbed by the press as the "Yankee Moslem," Webb became the face and voice of Islam in late nineteenth-century America. However, his flurry of activity endured only a few short years; the movement became moribund, hamstrung by an ambivalent society, lack of funds, internal division, charges of financial impropriety, and Webb's own race and class biases. He failed to arouse more than mere curiosity and supplied fodder for cartoons and editorials. By 1898 Webb had receded from his ambitious endeavor and lived out the remaining two decades of life in relative obscurity.

YANKEE MUSLIM, ED. BY BRENT D. SINGLETON

This work explores Webb's Indian travels and lectures prior to the launching of his Islamic mission to America. It provides readers with a corpus of primary material concerning the history and development of the first American Muslim movement, with special emphasis on its relationship to Indian supporters. When he wrote the travel journals, Webb meant for them to be mere unedited notes to be used in later writings, as a consequence they often lack the descriptive detail that readers of published travel journals may expect. Notwithstanding, the journals fill an important gap in the documentary evidence concerning Webb and his movement. The unedited documents allow historians to glimpse Webb's raw thoughts, personality, and beliefs. For instance, throughout the second half of the work, readers will note the transformation of Webb's worldview as his impatience and prejudices tear away at his vision of Asia and his reverence for Eastern culture and philosophies. Webb is left with an unenlightened, stereotype-ridden view of his Indian hosts, all of which illustrates a growing disconnection that foretells the collapse of his Islamic mission to America before it even begins.

Islam in America Prior to 1893

Some scholars have argued that Muslims were present in America prior to the voyage of Columbus, while others believe the first Muslims to arrive were remnants of the persecuted Moorish community who helped the explorer navigate to the New World. Regardless, the first well-documented person of Muslim origin to land in America was Estevan, an enslaved Black Moroccan. He was only one of three surviving members of Alvar Núñez Cabeza de Vaca's ill-fated exploration of the southern mainland of North America during the late 1520s and early 1530s. Significant numbers of Muslims did not arrive in America until at least a century later and like Estevan were also enslaved Africans.

The total number of African Muslims enslaved in the United States over the centuries is incalculable. However, it has been estimated that ten percent or more of the slave population of America may have been Muslim during the late eighteenth century.[1] These assessments rely mainly on records of the ethnicity and place of origin of likely Muslim captives. The historical record does include the stories of men such as Ayuba ibn Suleiman Diallo, Abd ar-Rahman Ibrahima, Bilali Muhammad, and Umar ibn Said, all literate Muslims who were robbed of their dignity in bondage and servitude, but retained at least some of their Islamic identity.[2] For each Muslim individually documented during slavery, there were thousands of Muslims who toiled in obscurity, growing evermore disconnected from their homeland and religion of birth. The harshness of American slavery and the forced Christianization of enslaved Africans destroyed nearly all vestiges of Islam. What little remained often returned back to Africa with the fortunate faithful or perished in America in the hearts of the unfortunate masses. Very few traces of Islam were passed on to generations beyond slavery, but Michael A. Gomez argues that the epicenter of post-slavery African Muslim identity was the coastal regions of Georgia and South Carolina. Here there were inklings of Islamic communities through the late nineteenth and early twentieth century.[3]

During the nineteenth century, there were also instances of Muslim traders and other travelers from various reaches of the Islamic world in America. Their presence was rarely documented except for a few cases such as famed camel breeder Hajj Ali, who worked for the U.S. Army during the late 1850s. During the second half of the nineteenth century and up to World War I, the number of immigrants from Muslim regions was rather paltry and predominantly made up of Syrian Christians and small numbers of Muslims from various sects.[4] In the mid-1890s, Hassan Ben Ali, an Arab-American Muslim planning to build a mosque in New York, estimated that 600 Muslims lived in the city at a time when the metropolis held roughly 2.5 million citizens. By

---

[1] Sulayman S. Nyang, *Islam in the United States of America*, 13.
[2] For a bibliography of African Muslims enslaved in America see Brent Singleton, "The Ummah Slowly Bled."
[3] Michael A. Gomez, *Black Crescent*, 159-162.
[4] Jane I. Smith, *Islam in America*, 51.

Ali's own admission his community was not interested in proselytizing, but rather in maintaining the Islamic virtues of visiting Muslim students.[5] In this milieu, Webb launched his audacious plan to bring Islam to American society at large.

Alexander Russell Webb's Childhood and Early Adult Life (1846-1873)

The Webbs were a well-established upstate New York family. Alexander Russell Webb, the second child of Alexander Nelson Webb and Caroline Elizabeth Lefferts, was born in Hudson, Columbia, New York, on November 9, 1846. The couple had relocated to Hudson from the Syracuse area in 1844 when Alexander Nelson was hired on as a printer for the *Columbia Washingtonian*, a local temperance newspaper. In the spring of 1847 he purchased the *Washingtonian* from Warren Rockwell and continued its publication while also founding the *Hudson Daily Star*, the first daily newspaper in the county. [6] Alexander Russell was just over a year old when his father embarked on the venture that would occupy the elder Webb for the next 35 years and leave an indelible impression on the child.

Hudson was a small riverfront city of less than 7,000 inhabitants in the midst of economic transition from whaling to shipping and was fast becoming a warehousing and railroad hub. [7] The Webb household at 338 Diamond St. was full of life and inhabitants. Residents over the years included Webb's parents, his five siblings, his Aunt Helen (his father's sister), several printing apprentices, and various servants. Twenty-two years separated Edward Cook and William Bunker the eldest and youngest Webb children; between them were Alexander Russell, Herbert Nelson, Carrie

---

[5] "Arabs in America," *Steubenville Herald* (OH), December 12, 1896, p. 7.

[6] Charles Williams Upton, *The History of the Hudson (New York) Daily Star, 1847-1940*, 15-17.

[7] Stephen B. Miller, *Historical Sketches of Hudson*, 106-7.

Lefferts, and Anna May.[8] Despite its small size Hudson provided many Twainesque amusements and diversions for children, from bullfrog fights and wheelbarrow racing to vaudeville troupes and dance contests. Webb may have even been a member of the Cadets of Temperance the children's branch of the Columbia County Temperance Society for which his father was secretary.[9]

Religion played a key role in the lives of the residents of Hudson; initially a Quaker town, it soon attracted Presbyterians, Methodists, Episcopalians, Baptists and others. Webb's father was a staunch Presbyterian, but it was said that "...aside from his devotion to temperance, which he supported vigorously throughout his entire career, he displayed no disposition to engage in crusades or indulge in fanaticism."[10] During Webb's childhood there were several revival movements in Hudson, particularly in 1855 and 1858, providing him with a glimpse of religious fervor.[11] Webb attended the Presbyterian Sunday School, yet never cared for what he deemed the "...abstruse discourses of the minister."[12] By his own admission he was a wild boy and went to the Presbyterian (and later) Episcopal churches solely for "...seeing nice-looking girls and escorting them home...I gave religion no thought."[13] Later, at the age of

---

[8] In 1855 the children's ages were listed as: E. Cook 14, Alexander R. 7, Hubert H. [sic] 3, and Carrie L. 6 mo. Alexander N. Webb household, 1855 New York State Census, Columbia County, Hudson, 4th ward, microfilm roll 30. North Salt Lake, UT: Heritage Quest; In 1865 Anna M. is listed as 6 and William B. as 4. Alexander N. Webb household, 1865 New York State Census, Columbia County, Hudson, 4th ward, microfilm roll 122, family 353. North Salt Lake, UT: Heritage Quest. Other residents of the household were listed in, Alexander N. Webb household, 1850 US Census, Columbia County, Hudson, 2nd ward, microfilm roll M432_491, page 230.
[9] Upton, *History of the Hudson Daily Star*, 94-96.
[10] Ibid., 18.
[11] Ibid., 72-3.
[12] Alexander Russell Webb, *Three Lectures*, 24.
[13] Alexander Russell Webb, *Lectures on Islam*, 30. For text of the Webb interview concerning this matter see Appendix A.

twenty he gave up on religion altogether and drifted into materialism for nine years.[14]

Webb's father was a vociferous advocate of both the public and private schools of Hudson and Webb attended several of these institutions.[15] At age fourteen, as the first shots of the Civil War rang out, Webb enrolled at Warner's Home School, a boarding school in Glendale, Massachusetts. He remained there for nearly three years before returning to Columbia County and entering Claverack Academy and Hudson River Institute.[16]

After completing his studies at Claverack, Webb began an apprenticeship in the jewelry store of Charles E. Butler located in the same building as his father's newspaper.[17] Here he learned watchmaking and jewelry sales, skills he would turn to often as a young man whenever he was in need of money. Despite having grown up at the foot of the printing press Webb chose not to enter his father's profession, even as the rightful heir to the business after his older brother

---

[14] Ibid., 19.

[15] In a biographical statement written to attain a consul position Webb states that he had attended two private schools in Hudson. However, a later biography, which he included in one of his own works states that he attended the public schools in Hudson. "Alex R. Webb Statement of Age, Education, Employments, etc." 1887, Letters of Application and Recommendation, 1885-1893, Record Group 59, 250/47/23/4 Box 131, National Archives and Records Administration (For text of statement, see Appendix B); L. Grebsonal, "The Mohammedan Propagandist," 204.

[16] Claverack Academy began as the Washington Seminary in 1777, and was chartered Claverack Academy and Hudson River institute in 1854. Despite often being called Claverack College, the institution was more on par with an advanced high school. Webb's father often extolled the virtues of Claverack Academy in the *Daily Star*, in 1862 he stated, "Everywhere throughout the state this institution is noted for discipline, high standard of education and economical administration. It is understood to be the model school of the state." Upton, *History of the Hudson Daily Star*, 80 and 154.

[17] "Alex R. Webb Statement of Age, Education, Employments."

chose to study medicine.[18] Instead, his younger brother Herbert apprenticed and inherited the newspaper business upon his father's retirement in 1873.

After nearly two-and-half years of unpaid work under Butler, and "Desiring to seek his fortune in the West..." Webb moved to Chicago in the late 1860s to work for Giles Bros. & Co., a large jewelry store.[19] Chicago was a rapidly growing city and a place where a young man with ambition could make a name for himself. On May 6, 1870, Alexander Russell married his first wife, Laura Cordelia Conger, and moved in with his in-laws, L.W. (Lucien West) and Elizabeth Conger.[20] Webb's new father-in-law was a partner in the Conger Bros. & Co. cattle brokerage at the Union Stock Yard.[21] Soon after his daughter's marriage, Conger quit the cattle business and staked his capital on his son-in-law's knowledge of the jewelry business, establishing Conger & Webb Co., a watchmaking and jewelry concern.[22] As was the case with many facets of Webb's life at this time his jew-

---

[18] Edward Cook, joined the 24th Regiment New York State Volunteers as a surgeon after he completed training at Albany Medical College. E. Cook later became Chief of Staff at the Homeopathic Hospital on Ward's Island, New York, before moving to San Francisco. In 1880 he set up a practice in Honolulu, Hawaii and over the next three years became a noted expert on leprosy. He was appointed to a number of government positions including, inspector of schools, and head of the insane asylum. He was known to frequent social events and attended medically to the Hawaiian royal family at Iolani Palace. In 1888, he returned to San Francisco and opened a dermatology practice. Upton, *History of the Hudson Daily Star*, 106; *Master Hands in the Affairs of the Pacific Coast*, 137; and Webb, E. Cook. A-1074. Physicians File. Mamiya Medical Heritage Center, Hawaii Medical Library.

[19] "Alex R. Webb Statement of Age, Education, Employments." Giles Bros. was located at 142 Lake. *Edward's Annual Directory of Chicago*, 1868-69.

[20] Illinois State Archives, *Illinois Statewide Marriage Index, 1763—1900*, http://www.sos.state.il.us/GenealogyMWeb/marrsrch.html [accessed 12/2/04]; and *Edward's Annual Directory of Chicago*, 1870.

[21] Ibid., 1868-69. The family lived at 73 26th St.

[22] The jewelry store was located 135 Clark St. Ibid., 1870.

elry business lacked permanency, perishing in the Great Chicago Fire of October 1871.[23]

Desiring a fresh start in familiar surroundings, Webb returned to his home state of New York, where his father had arranged a position for him as a jewelry salesman for Tiffany and Co. on Union Square, New York City.[24] In 1873, at the request of his wife's parents, Webb moved back to Chicago to alleviate his in-laws' concerns about Laura's health problems, which apparently arose during pregnancy.[25] Back in Chicago, Webb spent a short time employed at the jewelry house of Wendell & Hyman and also opened the Alexander R. Webb & Co. Billiard Hall.[26] On May 25, 1873, Webb's first child, Clarence Herbert, was born.[27] The family's second stay in Chicago was brief as Webb was given an opportunity that he could not pass up. His father-in-law, now a partner and cashier at the Putnam Bank in Unionville, Missouri had purchased an interest in the local paper, the *Republican*. He offered Webb the associate editor position as an enticement to move to the small town.[28]

Life in Unionville, Missouri (1874-1876)

Located in the north-central border region of Missouri and Iowa, Unionville was the seat of Putnam County. After living in large cities for more than five years Webb embraced the return to his bucolic roots. The family arrived in early January 1874 and Webb wasted little time before he

---

[23] "Alex R. Webb Statement of Age, Education, Employments."

[24] Ibid.

[25] Ibid.

[26] Ibid. The billiard hall was located at 758 Michigan Ave. and Webb lived at 72 Ellis Ave. *Edward's Annual Directory of Chicago*, 1873.

[27] This child is not mentioned in any further sources concerning Webb's children. Clarence Herbert Webb later took the last name of his stepfather, Reuben Hull Spicer Jr., after his mother remarried in 1879. Charles G.B. Conger, *A Record of Births, Marriages and Death of the Descendants of John Conger of Woodbridge N.J.*, 114.

[28] "Alex R. Webb Statement of Age, Education, Employments"; and *History of Adair, Sullivan, Putnam and Schuyler Counties, Missouri*, 562-64.

edited his first issue of the weekly Unionville *Republican* on January 8. He worked in conjunction with William O'Bryant on the paper and established himself in the community in a position not unlike the one his father held when he was a child. Soon Webb became politically active and civic-minded, joining many clubs and associations including: Unionville Temperance Association, Literary Society, Unionville Base Ball Club, State Board of Centennial Managers for Putnam County, Unionville Literary and Dramatic Association, and the Putnam County Republican Central Committee.[29]

Webb's most active political period was during the congressional election of 1874. He wrote scathing weekly editorials and articles for four months vilifying Republican incumbent Congressman Ira B. Hyde and his "...corrupt clique." Despite the misgivings of many Putnam County Republicans, the party's central committee decided to put their weight behind Hyde. This was intolerable to Webb who already had Democratic leanings. In response, Webb advised his readers to support the populist third-party People's Movement "...in its warfare against Bourbonism," a prudent move since he realized that openly supporting a Democrat would have raised the ire of local Republican officials.[30] Webb's partner, William O'Bryant, provides a candid description of the political fallout of the November 1874 congressional election:

> "Webb was a brilliant fellow, but somewhat erratic...An incident occurred while Webb was connected with the *Republican* which was of doubtful political consistency, but it seemed expedient at the time. The representative in congress, Mr. Hyde, in dispensing pa-

---

[29] "Town, County and State," *Republican* (Unionville), March 19, 1874, p. 3; Ibid., December 10, 1874, p. 3; Ibid., May 27, 1875, p. 3; Ibid., July 1, 1875, p. 3; Ibid., November 11, 1875, p. 3; Ibid., December 2, 1875, p. 2.

[30] "Warfare on Bourbonism," *Republican* (Unionville), July 30, 1874, p. 2; and "On the Wrong Tack," Ibid., July 29, 1875, p. 2.

tronage, had mortally offended the management of the *Republican*, to such an extent that the paper refused to support his for reelection. Judge DeBolt was the opposing candidate. The judge was always very popular in Putnam County, having been on the circuit bench there for years. The *Republican* did not hoist his name at the head, but gave him its support. DeBolt was elected by a good majority. The *Republican* was given the credit of making one congressman and unmaking another by the vote directly influenced by its action in this campaign. What ever of blame or credit for this, belongs to Webb, for he had full control of the paper at the time."[31]

Although victorious, Webb soured on politics after the rough-and-tumble campaign and lingering criticism from party-line Republicans. Speaking of himself in the third person, Webb described his situation at the time, "He had never taken more than a superficial interest in politics and soon found that his ideas were too strongly Democratic to admit his running a Republican paper to the satisfaction of the local leaders of the party..."[32]

Politics had lost its appeal, but Webb began to take a real interest in drama. He filled his free time with participating in dramatic readings and he was elected secretary of the Unionville Literary and Dramatic Association. The group's activities focused on Shakespeare and Webb read such roles

[31] Letter from Wm. T. O'Bryant. http://www.unionvillemo.com/letter.htm 12/6/1909 [accessed 12/4/04]. Webb gloated after the victory, writing the headlines: "The Hyde clique eternally and everlastingly used up. The people condemn rascality, and decide in favor of truth and honesty. Not a single clique man elected. Ira B. Hyde justly rebuked." *Republican* (Unionville), November 5, 1874, p. 2.

[32] "Alex R. Webb Statement of Age, Education, Employments."

as Hamlet, Brutus, and Shylock from the *Merchant of Venice*.[33]

In succeeding years, several co-owners and editors of the *Republican* followed William O'Bryant, who sold his interest in December 1874.[34] Webb's final partner was his brother Herbert who joined the *Republican* after moving his wife, parents and sister to Unionville in March of 1876. Herbert had been working for his father for many years and took over operations of the *Hudson Daily Star* in May 1873 when his father's health failed.[35] On March 16, 1876, the *Republican* was published under the management of the "Webb Brothers." Seemingly content with the new partnership and close proximity to his parents, Webb's world was about to be disrupted once again. Only four months after Herbert's arrival, L.W. Conger abruptly sold his interest in the *Republican* and Alexander Russell was forced to step down from his position. The last issue he edited was July 20, 1876. It appears that marital discord between Webb and his wife caused a rift that led to his dismissal and a petition for divorce. Webb quickly departed for St. Joseph, Missouri, leaving his family behind.[36] Although Webb remained in St. Joseph a mere three months, his employment at the *Gazette* introduced him to noted poet and newspaperman Eugene Field, who was city editor of the paper. Soon after Webb's arrival, Field moved to the editorial staff at the *Journal* in St. Louis and a few months later—likely at Field's behest—Webb was hired on as a beat reporter for the morning edition of the paper.[37]

---

[33] "Town, County and State," *Republican* (Unionville), November 11, 1875, p. 3; Ibid., January 27, 1876, p. 3; Ibid., February 3, 1876, p. 3; and Ibid., February 24, 1876, p. 3.

[34] Co-owners included, Clarence Conger (December 1874-March 1875) and O.J. Brown (March 1875-March 1876). *History of Adair, Sullivan, Putnam and Schuyler Counties*, 564.

[35] Upton, *History of the Hudson Daily Star*, 115-16.

[36] *History of Adair, Sullivan, Putnam and Schuyler Counties*, 564.

[37] Eugene Field, *The Complete Tribune Primer*, 3-4; and "Alexander R. Webb," *Evening Record* (Bergen County), October 3, 1916, p. 1.; and "Alex R. Webb Statement of Age, Education, Employments."

The St. Louis Years (1877-1887)

Webb had tried living the life of his father, but was ready for the more cosmopolitan environment St. Louis provided. By the late 1870s St. Louis was losing some of its preeminence to Chicago as the gateway to the west, but it was still a thriving and growing city. The next 11 years in St. Louis would significantly change the course of Webb's life, the first year being particularly eventful. On March 1, 1877, Webb's father died in Unionville. On June 27 Webb married his second wife, Ella G. Hotchkiss, a widow and mother of a seven year-old daughter, Bessie.[38] The family took up residence not far from the Mississippi River.[39]

Webb's employment was in dizzying flux during his first years in St. Louis. Having a family to care for meant he needed a steady salary, which the proprietors of the *Journal* failed to provide. He turned to the jewelry trade once again, joining the firm of Eugene Jaccard and Co. Shortly thereafter the *Journal* and the *Evening Dispatch* were consolidated into one ownership and Webb was convinced to return as city editor of both papers once the salary issue had been addressed. This arrangement continued until late 1878 when Joseph Pulitzer purchased the *Dispatch* and the *Journal* was consolidated with the *Times*. Webb became a reporter for the *Times-Journal*. In 1880 he accepted a position with the *Globe-Democrat*, but returned to the *Times* when the *Times-Journal* restructured. In 1881, the *Times* amalgamated with the *Republican* and Webb was left unemployed. As had become customary, he returned to the jewelry trade, this time with Mermod, Jaccard, & Co.[40]

---

[38] Upton, *History of the Hudson Daily Star*, 116. Ella was born on December 18, 1848 in Cincinnati, OH. Bessie (Elisabeth F. Hotchkiss) was born on May 24, 1870 in St. Louis. "List of American Citizens," *Despatches from U.S. Consuls in Manila, Philippine Islands, 1817-1899*; January 1, 1891, (National Archives Microfilm Publication T43, Reel 10), Records of the Department of State, Record Group 59.

[39] The address was 1719 Cass Ave. *Gould's St. Louis Directory,* 1877.

[40] "Alex R. Webb Statement of Age, Education, Employments."; and *Gould's St. Louis Directory*, 1877-1882.

In reminiscing, Webb's longtime friend and fellow newspaperman, William A. Kelsoe, described Webb during this period, "The most versatile man we had was perhaps Alex. Webb. He could sing and dance, act and make a good speech, and there wasn't a better all-round newspaper man on the paper…Webb was a practical printer, and when he couldn't find work in the writing end of a newspaper he would join the subs in the composing room. Twice he left *The Republic* to stage new plays and take theatrical companies over the country…"[41] On another occasion Kelsoe wrote: "…Webb was a born newspaper man…could "Stick type" in the composing room, and was ready any time to take a new show with a new theatrical star out on the road; all these things he did when reporting was dull in St. Louis, and in the reporters' class there were very few, if any, who could handle a big news story better than Webb or turn out copy faster than he could…"[42] Kelsoe alludes to Webb's moonlighting as a theatrical production manager. In 1879-80 Webb was an agent for the comedic duo, Robson and Crane, and successfully guided them through the show circuit.[43] That same year Webb's second son, Russell Lorenzo, was born and the family moved once again.[44] Webb then went on to manage several stage productions, which were generally panned by the critics.[45] In response, he returned to journalism in the later part of 1882 as editor of the *Dramatic Critic*, preferring to write rather than receive critical reviews.[46]

---

[41] "Alex. Webb Versatile," *St. Louis Republic*, July 12, 1908, part 15, 2.

[42] William A. Kelsoe, *St. Louis Reference Record*, 182.

[43] "Hopes to Islamize America," *Chicago Tribune*, December 21, 1892, p. 2. Stuart Robson, 1836-1903, and William Henry Crane, 1845-1928, presented the well-received act *The Comedy of Errors* during the 1879-1880 season. Donald Mullin, comp., *Victorian Actors and Actresses in Review*, 138.

[44] Russell Lorenzo Webb was born on May 24, 1879 and the family moved to 1218 Madison. "List of American Citizens"; and *Gould's St. Louis Directory*, 1880.

[45] "Hopes to Islamize America," *Chicago Tribune*, December 21, 1892, p. 2.

[46] The *Dramatic Critic* billed itself as the "only dramatic publication west of the Mississippi." The Dramatic Critic Company listed E.G. Webb

As had happened when Webb decided to settle down in Unionville, years of instability and wanderlust had given way to a yearning for permanence and meaning. On November 1, 1883, Webb returned to the daily newspaper business as a reporter for the *Republican*, his last employer in St. Louis. More significantly, after years of drifting deeply into materialism, Webb embarked on a spiritual quest. While in Unionville Webb had attended séances. His interest in spiritualism and Eastern religion was reenergized in early 1884 when he began to look seriously at Asian religious systems.[47] In March, he joined the newly-formed Pioneer Theosophical Society.[48]

The Theosophical Society was founded by William Q. Judge, Henry S. Olcott, and Helena P. Blavatsky in 1875 as a means of exploring the esoteric traditions of Buddhism. The group also investigated mystical and spiritual phenomena. The main precepts of the Theosophical Society were to "…unite in an effort to establish the Universal Brotherhood of Man as a fact, and influence others not members of the society to take up the study of theosophy, and under such instruction lead them to a practical knowledge of universal brotherhood."[49]

With his newfound religious outlook Webb's behavior changed accordingly. Webb had previously been considered a "…high liver, a skeptic, and man about town…" by some of his associates, who were shocked to see his transformation. He simplified his life. He gave up outward displays of wealth, ate a strictly vegetarian diet, and refused to

---

as an owner, most likely Webb's wife, Ella G. J. Thomas Scharf, *History of St. Louis City and County*, 957.

[47] "Town, County and State," *Republican* (Unionville), November 25, 1875, p. 3.

[48] "Alex R. Webb Statement of Age, Education, Employments."; and David Wietersten, Theosophical Society Archivist, e-mail message to editor, February 16, 2004. The Pioneer Theosophical Society formed in 1884 and dissolved in 1887; with two separate societies rising in its place, forming the Pranava and Arjuna branches. "In America," *Path*, April 1886, p. 31; Ibid., "In America," October 1887, p. 223.

[49] William Hyde and Howard L. Conrad, eds., *Encyclopedia of the History of St. Louis*, vol. 4, 2261.

indulge in alcohol, tobacco, or stimulants of any kind.[50] At about the same time as the birth of his first daughter, Mary Caroline, in 1885, Webb began a stint of celibacy, which lasted at least two years.[51] More and more of his free time was spent engaged in religious studies and conversations with others looking for the "Truth."

After several years of studying Buddhism and other Asian religions, Webb began to look seriously at Islam. In 1886 he initiated a correspondence with Mirza Ghulam Ahmad of Qadian, Punjab, India, who later founded the Ahmadiyya Movement.[52] Webb sought him out after reading a letter Ahmad had published in a newspaper inviting people to the religion of Islam and offering a reward of 10,000 rupees if anyone could refute his three hundred arguments supporting the authenticity of the Qur'an and prophethood of Muhammad.[53] The correspondence was cordial and at times reverential, and included literature about Islam written by Muslims. From the letters it is obvious that Webb truly wanted to be convinced that Islam was the right path and he claimed he only needed guidance from a Muslim to complete his journey. Although only partially convinced of the truths in Islam at the time, he was struck with the idea of an Islamic mission to America as early as 1887.[54]

The initial correspondence between the two men occurred prior to Mirza Ghulam Ahmad's controversial pro-

---

[50] "Hopes to Islamize America," *Chicago Tribune*, December 21, 1892, p. 2; and "Alex R. Webb Statement of Age, Education, Employments."

[51] Mary was born May 25, 1885. "List of American Citizens"; Reference to celibacy comes from Ghulam Ahmad letter. Ahmad, *Shahna-i-Haq*, 81-88. For text of correspondence between Webb and Ahmad, see Appendix C.

[52] Mirza Ghulam Ahmad, 1835-1908, in 1889 he claimed to be the Mujaddid (Renewer of the Faith) and later the Masih (promised Messiah). Despite these claims being seen as heretical by many Indian Muslims, the Ahmadiyya movement continued to grow. Richard C. Martin, ed., *Encyclopedia of Islam and the Muslim World*, 32.

[53] Basharat Ahmad, *Mujaddid Azam*, 157.

[54] Ahmad, *Shahna-i-Haq*, 81-88. For text of correspondence between Webb and Ahmad, see Appendix C.

23

nouncements of being the promised Messiah in 1891.[55] Although Webb was a Theosophist, a firm believer in mysticism, and a non-sectarian, his contemporaries saw his conversion to Islam as having been in the vein of orthodox Sunni Islam. In South Asia, the Sunni majority has long castigated the followers of Ghulam Ahamd as heretics, accusing them of not upholding the Muslim tenet that Muhammad is

---

[55] While in India Webb reportedly told his associate, Moulvi Hassan Ali, "I am very grateful to Mirza Ghulam Ahmad. It is because of him that I have had the honor to join Islam. I would like to meet him." However, because of Mirza Ghulam Ahmad's negative image in India at the time it was decided that he would forgo meetings with Ahmad or his followers, so as not to disrupt fundraising. While in Lahore, Webb was asked as to why he did not go to Qadian to visit Ahmad and his wry reply was "What is there in Qadian?" Webb later regretted not meeting with Mirza Ghulam Ahmad and despite his curt remark, remained in correspondence with him until near Ahmad's death in 1908. In 1906, a letter from Webb to Ahmad reportedly stated, "Alas! I came to India but did not visit you, although it was through you that I found the right guidance. In not meeting you, I tried to please some people so that they would give donations. They did not keep their promises and no donations were made. Now I regret greatly that I deprived myself of meeting a man of God for such people." (Ahmad, *Mujaddid Azam* 159-161) It must be stated that none of the above conversations nor Mirza Ghulam Ahmad are mentioned in Webb's Indian travel journals. Webb continued to read Ahmadi literature for several decades after his correspondence began. In 1915, the *Review of Religions*, the main organ of the Ahmadiyya Movement, stated that Webb had been in continuous correspondence with the editors since at least 1893. He even provided rather flattering reviews of the publication. (Maulana Dost Mohammed Shahid, "Review of Religions," 21-23). Despite all of his contacts with Mirza Ghulam Ahmad and his followers it seems that Webb never joined the Ahmadi movement. This sort of tolerance is not an uncommon phenomenon amongst converts to Islam who are often grateful to whom ever initially leads them to the religion. It occurred many decades later when the mass of African-American Muslims converted to orthodox Islam in the 1970s after the Nation of Islam (NOI) leader Elijah Muhammad died. They took to orthodox beliefs and practices, and despite rejecting the beliefs of their former leader still felt gratitude to Elijah Muhammad for leading them to Islam.

the seal of the prophets. The Ahmadiyya have been persecuted for their beliefs since the 1890s, and despite affection
for Ahmad, Webb was careful to stay clear of him and his
followers when he visited India in 1892.

Webb remained a newspaperman throughout this
time period, being named assistant city editor at the *Republican* in 1886.[56] He also had access to a large collection of religious books numbering around 13,000 volumes and spent
four to five hours a day in study, yet, he felt constrained
from pursuing his religious studies in a fashion suitable to
his needs.[57] He wanted to be among Asians and see their
faith in action and ask them questions. Therefore, in the
spring of 1887, with the assistance of influential friends,
Webb lobbied President Grover Cleveland's administration
for an appointment as a U.S. consul in Asia. He specifically
wanted to be sent to Bombay or Singapore, but was willing
to be sent anywhere in India. Those who petitioned in his
favor included the governors of Missouri and Kentucky, the
mayor of St. Louis, a U.S. senator, several congressmen and
many past employers, friends, and associates.[58] There were
no openings for the cities he requested. Instead he was offered the position of consul at Manila, Philippine Islands,
which he accepted on September 26, 1887.[59]

American Consul at Manila, Philippine Islands (1888-1892)

Before Webb had even finished packing for Manila,
he put forth a request for a transfer to Calcutta or Singapore
should an opening arise.[60] On November 1, 1887, the Webb
family departed for San Francisco and a week later boarded

---

[56] *Gould's St. Louis Directory,* 1886.

[57] Webb, *Lectures on Islam,* 19.

[58] "Letters of Application and Recommendation, 1885-1893." Record
Group 59, 250/47/23/4 Box 131, National Archives and Records Administration.

[59] Letter from Alex R. Webb to Thomas F. Bayard, 26 September 1887.
Ibid.

[60] Letter from Alex R. Webb to Thomas F. Bayard, 12 October 12 1887.
Ibid.

the steamship *Gaelic*.[61] The month long voyage across the Pacific included scenic stops in Honolulu, Yokohama, and Hong Kong before arriving in Manila on December 10th. On January 1, 1888, Webb assumed the American consulship from Julius G. Voigt. His first acts were to move the consular office and hire his step-daughter, Bessie, as clerk.[62] Soon after Webb's arrival his third child, Nala, was born.[63]

The next four-and-a-half years would be spent fulfilling his duties as the American consul and reporting to his superiors in Washington. The correspondence was primarily mundane reports on agriculture, trade, and economics, and acknowledgement of receipt of orders from the assistant secretary of state. He often reported as well on destitute and abandoned American seamen, shipwrecks involving American ships, and deaths of American citizens. Webb socialized with the American and British expatriate communities as well as foreign diplomatic circles. In November 1888, it was suggested by his long-time friend, David B. Sickels, that Webb be transferred to Bangkok.[64] However, this was never acted upon, and a year later in the fall of 1889 a letter writing campaign began in earnest to get Webb transferred to Singapore.[65] Unfortunately for Webb, a new Republican admini-

---

[61] *Despatches from U.S. Consuls in Manila*, Reel 8, November 1, 1887. The *Gaelic* was of the Occidental & Oriental Steamship Co. For an account of a voyage from San Francisco to Hong Kong on the *Gaelic*, see Karl Irving Faust, *Campaigning in the Philippines*, 289-300.

[62] Before Voigt was able to leave Manila he died of dysentery. *Despatches from U.S. Consuls in Manila*, Reel 9, April 17, 1888. The consul office was relocated to No. 15 Plaza de Santa Ana Sampaloc. Ibid., January 1, 1888.

[63] Nala D. Webb was born February 11, 1888. "List of American Citizens."

[64] Letter from David B. Sickels to Thomas F. Bayard, 8 November 1888. "Letters of Application and Recommendation." David B. Sickels, 1837-1918, was Consul to Siam from 1876-1881 and co-founded the American Surety Company. "David Banks Sickels Dies," *New York Times*, July 20, 1918, p. 9.

[65] Several letters from Sickles and other prominent citizens in St. Louis requested the transfer. Ibid.

stration was now in place and the Democratic machine that had secured his appointment no longer held sway in Washington.

When not involved in his official duties, Webb investigated Islam, and read the works of mainly Muslim Indian authors. In describing Webb prior to his departure for Manila an associate said, "When he left here…he was perhaps more of a Buddhist than a Moslem, though his Buddhism might not have been recognized by the ordinary follower of Buddha."[66] Nonetheless, Webb was now fully committed to being a Muslim and in 1889 he met a Parsee man on business from Bombay who introduced him to Budruddin Abdullah Kur, and a correspondence pertaining to Quranic doctrinal matters ensued.[67] In July 1891, Webb formally requested a transfer from Manila himself, and suggested a post in Singapore, Colombo, Calcutta, Bangkok, Tangier, Beirut, Jerusalem, or Constantinople, among others. He cited the failing health of his wife and one child (due to the climate of the Philippines), as well an increase in duties on goods that made the cost of living unbearable for a large family on a salary of $2,000 per annum. However, due to the similar climates and costs of living of some of the requested transfer cities the real impetus for wanting to leave Manila appeared to be his religious studies. In the transfer request Webb further stated, "…my original purpose in coming to the East, was to pursue, during my leisure time, the study of the Oriental religions, but am in a post where the Catholic church is supreme and where no other form of religion is allowed a representation. For this purpose, therefore, I could not be more unfavorably situated."[68]

While Webb waited nearly a year for a response, a more compelling opportunity arose. Budruddin Abdullah Kur had published several of Webb's more interesting letters

---

[66] "The Land of Pagodas," *St. Louis Republic*, December 4, 1892, p. 10.

[67] Webb, *Three Lectures*, 4; and Alexander Russell Webb, "Preaching Islamism," 469.

[68] *Despatches from U.S. Consuls in Manila*, Reel 10, July 26, 1891; Salary taken from: Albert T. Volwiler, *The Correspondence Between Benjamin Harrison and James G. Blaine, 1882-1893*, 265.

in various Bombay newspapers. These caught the attention of wealthy Memon merchant, Hajee Abdulla Arab, who had been buoyed by reading of a successful Islamic propagation movement in England and thought that a similar effort ought be made in America. Arab immediately sent a letter to Webb and received an enthusiastic reply to a proposed mission to the United States. In March 1892, Arab and Moulvi Serajuddin Ahmed visited Webb in Manila to discuss a suitable arrangement.[69] The exact date that Webb officially converted to Islam is unclear, but by this time Webb believed strongly enough in the tenets of Islam to gain the support of Hajee Abdulla.[70]

The American Islamic Propaganda, as Webb's mission became known, was no small matter philosophically or financially; the agreement between Webb and Hajee Abdulla was written into a contract and signed by both parties. Webb agreed to a salary of $200 per month for the duration of the three-year contract. The sum of $13,500 was earmarked for the formation and maintenance of a publication department and lecture course for year one with $10,000 for each of the following two years. [71] Webb also agreed to travel to India, Egypt, Turkey, and Liverpool, England to garner subscriptions for the mission, meet with members of the Indian Muslim community, and give lectures.[72] Webb found this additional proposition very attractive, allowing him to visit the land he had come to revere from his religious and literary studies. Upon returning to India, Hajee Abdulla Arab and

---

[69] Ahmad, *Mujaddid Azam*, 158; and Webb, "Preaching Islamism," 468-69.

[70] In an interview in 1892 Webb said he had "...embraced..." Islam five years prior, roughly at the time he was in correspondence with Mirza Ghulam Ahmad. Webb, *Lectures on Islam*, 31. Also, in an article that same year Webb wrote, "A course of investigation extending over a period of nearly three years led me to the positive conviction that Mohammed was all that he ever claimed to be..." Alexander Russell Webb, *Islam and Theosophy*, 425.

[71] "To My Oriental Brothers," *Moslem World and Voice of Islam*, January 1895, p. 3-4.

[72] Webb, *Lectures on Islam*, 18.

Muslim missionary, Moulvi Hasan Ali, began collecting donations and immediately raised 6,000 rupees in Hyderabad. In June a meeting was held at the Ismail Habib Mosque in Bombay to form a committee to support the propagation effort in America.[73] Based upon this information, Hajee Abdulla Arab in a premature fit of enthusiasm, sent a telegram to Webb stating that he was to proceed with the mission and it was time for him to resign his consul position.[74] This was a fateful decision. The groundwork had not been fully laid for the mission's funding, which resulted in major unforeseen financial implications in later years.

Regardless, Webb tendered his resignation on June 22, 1892, requesting a replacement post-haste on the innocuous grounds of having "...entered into an agreement to engage in business in New York City."[75] By August, Webb was given authority to turn over consul operations to William A. Daland and did so on September 5th.[76] That same day, the first report of his conversion to Islam appeared in American newspapers.[77]

Travels in Asia (1892)

After nearly five years in Manila, Webb began his journey to India, a locale he had long ago idealized and dreamed about visiting. The first leg of his trip was a wearying weeklong voyage though the weather-battered Sulu and China seas on an inadequately managed Spanish ship. Upon arriving in Singapore on September 14, Webb was struck by the "oriental" feel of the city; despite spending a half-decade in the Philippines he had never felt like he was in Asia. The weeklong stay in Singapore was originally intended as an opportunity to reacquaint himself with Hajee Abdulla Arab and to prepare for the Indian tour. However, Hajee Abdulla

---

[73] "The Propagation of Islam in America," *Times of India*, October 24, 1892, p. 5.

[74] Ahmad, *Mujaddid Azam*, 159.

[75] *Despatches from U.S. Consuls in Manila,* Reel 10, June 22, 1892.

[76] Ibid., July 7, 1892; and Ibid., September 5, 1892

[77] "News Briefs," *Trenton Times*, September 5, 1892, p. 6.

was out of town. Undeterred, Webb took advantage of the unexpected leisure time to do some sightseeing and write detailed entries in his journal.

In Singapore, Webb met with several important officials and local peoples and saw the local sites, including a number of Chinese social events and festivals. Webb visited his first mosque and ironically was unceremoniously turned away. He eventually gained entry to a smaller mosque, a rather untraditional congregation, but warm and welcoming nonetheless. Webb appeared to enjoy his time in Singapore and looked forward to the next leg of his journey which would take him to Rangoon after a brief stop in Penang.

Upon reaching the harbor of Rangoon on September 28, Webb was welcomed by hundreds of cheering Muslims who hoped to glimpse or touch the American Muslim. This was his introduction to life in a Muslim community. He enjoyed the brotherhood of his new faith; the Friday prayers at the main Mosque left him in awe. Webb gave several short speeches and one lengthy lecture on Islam and the proposed mission and explored the sights of Rangoon, perused local shops and markets, and watched elephants at work. Webb was reunited with Hajee Abdulla Arab and after eleven days the pair embarked on the long-awaited voyage across the Bay of Bengal for India.

When Webb first set eyes on the Indian coastline on October 12th his dreams of visiting the ancient land were finally realized. Once again, he was warmly welcomed by the local Muslims of Calcutta. He spent his time there visiting the local attractions as much as possible and meeting with local Muslim leaders. During his stay, Webb began to feel apprehensive about the reliability of his Indian colleagues, a foreshadowing of frustrations to come. After delivering a lecture and fulfilling his obligations, Webb and Arab traveled to Bombay by way of Patna and Benares, and visited the temples of the latter.

Webb's reception in Bombay was as cordial and well attended as those of the other cities; it was becoming all too commonplace and wearisome for him. He met his long-time correspondent, Budruddin Abdulla Kur, as well as Moulvi Hasan Ali who would travel with Webb and Arab for the remainder of the trip. His stay in Bombay lasted three weeks and was filled with the requisite sightseeing, meetings with local dignitaries, and speaking engagements. Webb's ill temper and impatience with local customs and British colonial governance began to fester while in Bombay. The trio of travelers left for Hyderabad on November 17th, stopped for two days in Poona and then traveled through the Nizam's Dominions.

In Hyderabad Webb settled in rather quickly and began his rounds of meetings and sightseeing. He was invited to many of the grand palaces of the ruling elites for dinners and receptions. It was thought that he would have a meeting with the Nizam as well, but to Hajee Abdulla and Hasan Ali's disappointment this could not be arranged. Webb delivered a lecture to a large audience in the Public Garden and prepared to travel on to Madras. However, much to Webb's dismay there were attempts to delay his departure for several days and his disgust with his Indian hosts reached its peak. Webb left Hyderabad alone on December 8th, needing a break from his travel companions. He spent an uncomfortable night in Wadi before starting for Madras.

Once in Madras, Webb seemed to finally enjoy his travels once again. His first order of business was to visit the headquarters of the Theosophical Society in Adyar where he met Col. Henry Olcott and other officials of the group. He then met with local Muslim leaders and made two public appearances—a lecture and the dedication of an asylum for Muslim converts. He stayed in Madras for only four days before traveling through the Nizam's Dominions to Agra, where he arrived on December 15th. Webb's surviving travel journals end abruptly in Agra, but he remained in India for an additional two weeks, visited Lahore and delivered a lecture similar to the one he gave in Bombay.[78]

Despite his many frustrations, and possibly out of longing for being among other Muslims, Webb later wrote, "The four months I spent in India form the pleasantest memory of my life. I lived among Mohammedans all of this time.

---

[78] Webb, *Lectures on Islam*, 15.

I found a richer, better, higher civilization among these so-called 'barbarians' than I can find in the great metropolis of the new world to-day."[79] He departed India on December 31, 1892 from Bombay to a crowd of well-wishers and set off to return to America with planned stops in Egypt and Turkey before reaching Liverpool, England.[80] Accounts of his time in these countries are absent from the historical record.

Islamic Mission to America (1893-1897)

On February 16, 1893, Webb arrived in New York aboard the steamship *Majestic* and quickly took up residence at 40 West 29th Street. He spent the first week regaining his bearings in the country he had not seen in over five years.[81] A week later, news of his arrival broke across the nation, Webb briefly describes the media reaction at the time: "...I found the object of my mission had been widely announced in newspapers. Many treated the unheard of advent of a Mohammedan missionary in America as a huge joke and as nothing more than a subject for the wit of the paragrapher and the cartoonist."[82] Of this period he also noted: "It was reported as soon as I landed in America that I had millions at my command to convert Americans, and at once the sharks began to flock to me from all directions."[83] An article in the *New York Times* erroneously stated that Webb had $150,000 to construct a mosque and others proclaimed he had an endless supply of rupees to convert Americans.[84]

All of the uninformed media attention was a distraction to Webb who had a specific set of goals for the propa-

[79] Webb, "Preaching Islamism," 469.

[80] "Muhammed Webb's Mission," *New York Times*, February 25, 1893, p. 1. It was discussed in a letter to the editor of *The Times of India* dated November 16, 1892 that Webb was to visit Turkey and Egypt. Webb, *Lectures on Islam*, 18.

[81] "The Yankee Mahometan," *New York Daily Tribune*, February 26, 1893.

[82] Webb, "Preaching Islamism," 468.

[83] "Fall of Islam in America," *New York Times*, December 1, 1895, p. 21.

[84] "Muhammed Webb's Mission," Ibid., February 25, 1893, p. 1.

ganda mission, which was geared towards enlightening Americans and extinguishing the prejudices held against Islam. Webb explains the movement's goals:

> "The American Islamic Propaganda is to be purely educational, although Mohammedan missionaries will come here and preach in various parts of the country when their services are required. But for the present efforts of all engage in the work will be to teach the intelligent masses who and what Mohammed was and what he really taught, and to overturn the fabric of falsehood and error that prejudiced and ignorant writers have been constructing and supporting for centuries against Islam. No one realizes the magnitude of this effort more fully than I do; but I have full confidence in the intelligence and justice of the American people, and in their willingness to give a fair and impartial hearing to any claim that may be properly presented to them."[85]

Furthermore, the stated goals were to establish a high-quality weekly journal, a free lecture room, library, and a book-publishing venture.[86]

The mission was organized on secular practical grounds as well as on some philosophical tenets of Islam. The overarching movement was deemed the American Islamic Propaganda. It was made up of formal operations such as the mission's headquarters, the Moslem World Publishing Company located at 458 West 20th Street in New York, and a loose confederation of Islamic study circles formed the American Moslem Brotherhood. The Moslem World Publishing Co. was established only a few months after Webb's arrival in New York, but it took until October 1893 for the

---

[85] Webb, *Islam in America*, 67.

[86] Ibid., 67-68.

Headquarters to officially open its lecture room and library. Weekly lectures were given on Friday evenings; on Sunday afternoons informal gatherings for conversation and questions were held. [87]

The American Moslem Brotherhood was made up of independent societies in various cities, which enabled people to study and discuss Islamic history, literature, and doctrine among themselves. In true Theosophical style one need not profess a conversion to Islam to be a member and individuals were encouraged to follow any path their studies led them on, to use their own judgment, and to believe only those doctrines which were in "…harmony with commonsense."[88] The independent societies would set their own dues—if they chose to have them—select a name for the group, and elect their own president, vice-president, secretaries, treasurer, and librarian. This information was sent to Budrudin Abdulla Kur, secretary of the Indian committee, for purposes of chartering each society.[89] Although each society was independent they were encouraged to register with the central office in New York for the purposes of facilitating communication and distribution of literature.

Webb envisioned study circles in every city and town, eventually evolving into Muslim communities, which by pure force of example would stamp out immorality in America.[90] However, only a small number of circles were chartered, among them Mecca Circle No. 1 and two others in New York City, Capital Circle No. 4 in Washington D.C., Khadijah Circle No. 6 in Brooklyn, and one each in Pueblo, Colorado, Woodcliff, New Jersey, and Baltimore.[91] Webb published few details concerning the activities of the various circles, but he did claim that many of the members were doctors, lawyers, and other intellectuals. One of the more active

---

[87] "Our Lecture Room," *Moslem World*, November, 1893, p. 11.

[88] "The Islamic Propaganda," Ibid., May, 1893, p. 11.

[89] "The Moslem Brotherhood," Ibid.

[90] Ibid.

[91] Ives, "The Story of the First American Convert to Islam," 7; Pueblo Colorado - Webb, "Preaching Islamism," 468; Baltimore - "More Light is Needed," *Moslem World*, September 1893, p. 16.

members was Albert Leighton Rawson, organizer of several of the study circles and well known within Freemasonry as a founding member of the Nobles of the Mystic Shrine, or Shriners. Regardless, few if any of the members, including Rawson, ever converted to Islam.[92]

Webb began publishing in conjunction with The Oriental Publishing Company soon after setting up shop in New York.[93] The first work published in 1893, *Islam in America*, was based upon some of the lectures he had given in India with additional editing and commentary. Later that year when his own Moslem World Publishing Company was able to handle books it published *A Guide to Namaz: A Detailed Exposition of the Moslem Order of Ablutions and Prayer*. The Moslem World Publishing Company published several books and advertised over one hundred other books for sale, but its main project was producing the newspaper *The Moslem World*, the official organ of the American Islamic Propaganda. The first issue of *The Moslem World* appeared on May 12, 1893. The original plan was to publish the 16-page paper monthly until October and change to a weekly schedule distributing the paper on Fridays.[94] As promised the paper received the highest quality printing, a fact noted even by the harshest critics of its content. Each issue was wrapped in a jacket with an ornately illustrated title header and a different Mosque featured monthly.

The newspaper consisted of articles written by prominent Western intellectuals and Muslims from around the world concerning various aspects of Islamic doctrine and history, such as a serialized version of Moulvi Cheragh Ali's *A Critical Exposition of the Popular Jihad*. There were also a plethora of news items from around the Islamic world. Some

---

[92] K. Paul Johnson, *Masters Revealed*, 25-30.

[93] The Oriental Publishing Company was located at 1122 Broadway, New York City. Webb was forced to use their services because Hajee Abdulla Arab was delayed in sending the first installment of $2,000 to begin the mission's own publishing company, which began operations on April 1, 1893. "To My Oriental Brothers," *Moslem World and Voice of Islam*, January 1895, p. 4.

[94] "Special Notice," *Moslem World*, June, 1893, p. 8.

articles were reprinted from foreign newspapers, but a significant portion was provided expressly for *The Moslem World* by various foreign correspondents. Webb included American newspaper articles, which both supported and reviled the Propaganda, using one to encourage his movement and the other as evidence of his persecution in the press. Oddly, the *Moslem World* included very little news about the progress of the American Islamic Propaganda itself, leaving it to be reported in the mainstream press.

The Moslem World lasted for only seven monthly issues, May through November 1893. As early as July of that year it was apparent that financial problems were arising, the price per issue was doubled from 5¢ to 10¢ with the explanation of, "This course has been found necessary in view of the cost of printing and engraving which is greater than was first estimated, and to cover the expense of sending large numbers of copies to poor persons who want the journal but are unable to pay for it." Furthermore, the notice explained that a considerable loss was expected to be incurred the first three years of publishing the paper.[95] The *Moslem World* could not be sustained past November.

Many problems arose concerning the overall governance of the American Islamic Propaganda in the succeeding months, but by June 1894 a much smaller publication entitled *The Voice of Islam* began publication. Although actually issued monthly, it was intended to be for weekly news updates with the larger sixteen-page *Moslem World* serving as the monthly newspaper.[96] Unfortunately, even the smaller four-page format at 3¢ an issue failed rather quickly, folding after three months in August 1894. The final journal published for the American Islamic Propaganda was a combined publication called *The Moslem World and Voice of Islam*. It used the smaller four-page format and was only intended to be a stop-gap measure until funding to produce the two separate publications in their original formats could be secured.[97]

---

[95] "Special Notice," Ibid., July, 1893, p. 8.

[96] "Our Plans," *Moslem World and Voice of Islam*, January 1895, p. 2.

    The author has not been able to locate copies of *The Voice of Islam*.

[97] Ibid.

The publication consisted almost exclusively of tidbits of news from the India, Turkey, and Egypt with a few letters from elsewhere, and news of the American Islamic Propaganda. The combined journal was easily the longest lasting, although it is unclear exactly when it ceased, it was published until at least February 1896.

Speaking engagements were a major part of the propaganda effort and began soon after Webb arrived in New York in February 1893. The first events were mainly invitation only receptions given in his honor, but he spoke to a large audience at the Aryan Theosophical Society in New York on March 7th, which was well received.[98] These initial events were all held locally while Webb busily built the organizational structure for his movement and worked on other projects. Webb remained largely out of the press for several months until he aroused national curiosity in May with his search to procure large tracts of land in eastern Alabama, Augusta, Georgia, and Jacksonville and Pensacola, Florida. He secured options to purchase 25,000 acres in Georgia and seven other equally large parcels to establish Muslim colonies for immigrants from India. He also looked into purchasing land in Chiapas, Mexico for similar purposes.[99] Although none of this work bore fruit the national press were amused at the prospect of multitudes of Muslims settling in the southern United States.

By the fall of 1893, with much of the organizational structure of the mission in place, Webb was ready to engage in a national speaking tour. In September, Webb was the only Muslim slated to speak at the prestigious World Parliament of Religions in Chicago, given in conjunction with the

---

[98] Samuel Brown of Jersey City, NJ gave Webb a reception on February 24th and Webb's old St. Louis friend, David B. Sickles, gave another in New York on February 27th. "Muhammed Webb's Mission," *New York Times*, February 25, 1893, p. 1; and "Mohammed Webb Talks to Theosophists," *New York Daily Tribune*, March 6, 1893, p. 3.

[99] "Colonies May Come Here," *Washington Post*, May 29, 1893, p. 1; and "Mohmmedanism in Mexico," *Arizona Republican*, August 27, 1893, p. 8.

World's Fair.[100] He gave two lectures—"The Spirit of Islam" and "The Influence of Islam on Social Conditions"—both of which were well attended.[101] He created a sensation when his first lecture was interrupted by some audience members who demonstratively objected to Webb's defense of some forms of polygamy. The *Chicago Tribune* described the interchange as follows:

> "Now I want to say to you honestly and fairly that polygamy never was and is not a part of the Islamic system. To engraft polygamy upon our social system in the condition in which it is today would be a curse. There are parts of the East where it is practiced. There are conditions under which it is beneficial [Cries of "No," hisses: and slight applause] But we must first understand what it really means to the Mussulman, not what it means to the American. I say that a pure-minded man can be a polygamist and be a perfect Christian [cries of "No," hisses and groans], but he must not be a sensualist. When you understand what the Mussualman means by polygamy, what he means by taking two or three wives, any man who is honest and faithful and pure-minded will say "God speed him." [Cries of "No," "Shame," hisses and applause.][102]

---

[100] Originally, in conjunction with supporting Webb's mission the propaganda committee in Bombay had planned to send several learned Moulvis to Chicago, however it never panned out. "The Propagation of Islam in America," *Times of India*, October 24, 1892, p. 5.

[101] For text of the two lectures see Barrows, John Henry, ed., *The World's Parliament of Religions; An Illustrated and Popular Story of the World's First Parliament of Religions, Held in Chicago in Connection with the Columbian Exposition of 1893*. Chicago: The Parliament Publishing Company, 1893, 989-996 and 1046-1052.

[102] "Defends Islamic Polygamy," *Chicago Tribune*, September 21, 1893, p. 9.

Instead of continuing the course of his lecture, Webb moved on to other topics in order to get his greater message across to those in attendance. Although this rebuke by some audience members received media attention nationwide, there were many parts of his lecture that were applauded and cheered by those assembled. After the lecture, Rev. George E. Post came forward and attacked Webb's views, reading aloud and partially misinterpreting several verses of the Qur'an. This was the only direct attack on any religion or participant at the Parliament.[103]

Undaunted, and perhaps buoyed by the abundant attention in the national press, Webb returned to New York and gave a succession of eight local lectures in a two-week period.[104] This appeared to be in preparation for an extensive national tour arranged by the Oriental Literary Bureau, whereby Webb would lecture in large cities and small towns across the country. From October 17, 1893 through March 5, 1894 he was scheduled to give over 40 lectures in cities in nineteen states throughout the midwest, east and south.[105] It is doubtful that Webb was able to speak at any of these venues because he canceled his agreement with the Oriental Literary Bureau soon after the tour's announcement. Instead he spoke at the Unitarian church in Ithaca, New York and planned an engagement in Rochester, New York, and a shorter tour to Chicago; St. Louis, Kansas City, and St. Jo-

---

[103] "Islam is Preached," Ibid., September 21, 1893, p. 9.

[104] Webb gave his lecture "The Faith of Islam" at: Chickering Hall, New York, September 29; YMCA, Brooklyn, September 30; Lee Avenue Academy of Music, Brooklyn, October 1; Association Hall, Trenton, NJ, October 2.; and five others cities. "Advertisement," *New York Times*, September 24, 1893, p. 7; "Coming Events," *Brooklyn Eagle*, September 30, 1893, p. 5; "Advertisement," *Trenton Times*, September 27, 1893, p. 8; and "Headquarters Opened," *Moslem World*, October 1893, p. 9.

[105] The itinerary of the national tour was for the following states: Ohio, Indiana, Michigan, Illinois, Iowa, Nebraska, Missouri, Arkansas, Kansas, Pennsylvania, Connecticut, Maryland, North and South Carolina, Alabama, Tennessee, Mississippi, Kentucky, and Texas. "Advertisement," *Moslem World*, October 1893, front matter.

seph, Missouri; and Grand Rapids and Ann Arbor, Michigan. He also reduced his total speaking events to four per month outside of his Friday lectures at the American Islamic Propaganda headquarters.[106]

The cancellation of the extensive tour with the Oriental Literary Bureau was due to financial problems; by November 1893, the American Islamic Propaganda was in turmoil. Webb could not afford the expense of either a protracted tour or being away from his headquarters for four months. Hajee Abdulla Arab sent funds in November, but was forced to cease further transfers when the rupee fell in value and his Indian counterparts reneged on their promised donations. The tour and *The Moslem World* were the first casualties; the latter ceased publication that month. Appraising the situation of late 1893, Webb later wrote, "But from loans and contributions from others than Hajee Abdulla Arab I should have been compelled to close my New York office on the first of January, 1894."[107]

Just as news of possible insolvency reached Webb, internal dissension arose within the American Islamic Propaganda. In December 1893, two former confederates, Emin L. Nabakoff and John A. Lant, broke away to form a rival Muslim group called the First Society for the Study of Islam, headquartered at No. 8 Union Square, New York City.[108] According to Webb, personal enmity between he and Lant caused the split. Webb reported that he became distrustful of Lant when it was revealed that he had a prison record and

---

[106] "Mr. Webb's Lectures," *Moslem World*, November 1893, p. 9.

[107] "To My Oriental Brothers," *Moslem World and Voice of Islam*, January 1895, p. 4.

[108] "Preaching Islam Here," *New York Daily Tribune*, December 11, 1893, p. 8. Emin L. Nabakoff was a native of Russia and moved to Alaska becoming a naturalized U.S. citizen when that region was purchased from Russia. He arrived in New York in August 1893 and immediately associated himself with Webb. "Personals," *Moslem World*, September 1893, p. 9; and "A Fourth-Floor Mosque," *New York Times*, February 4, 1895, p. 8; "Muezzin's Call Not Silenced," Ibid., December 18, 1893, p. 1; "Mohammed a Bone of Contention," Ibid., January 22, 1894, p. 5; and "Annual Meeting of the Liverpool Moslem Institute," *Crescent*, August 12, 1893, p. 239.

quickly expelled Lant from the headquarters. Lant was in-
credulous and believed Webb's objection on the matter was
disingenuous because his imprisonment had been politically
motivated. In the 1870s, Lant had run afoul of Anthony
Comstock and his controversial decency laws for publishing
and distributing literature with references to religious philosophies Comstock construed as antithetical to Christian
beliefs. Comstock dogged Lant for decades and had him arrested several times for misuse of U.S. mails.[109]

Nabakoff's disagreement with Webb appears to have
been motivated by religious differences based on Webb's
failure to establish formal Islamic practices in America. Nabakoff had been an active member of the Liverpool Moslem
Institute, a vibrant community of Muslim converts in Liverpool, England, which upheld many of the practices of orthodox Islam. The First Society claimed that it represented a
truer form of Islamic practice, using the Muslim call to
prayer, performed publicly in Arabic, as a central part of its
meetings.[110]

Muslims both home and abroad were maligning
Webb. Lant joined forces with Hamid Snow, a newspaper
editor and Muslim convert in India. While Snow castigated
Webb in the Indian press, Lant started publishing a newspaper called *The American Moslem*, in which he lambasted
Webb as a charlatan who had misused his funds from India.
Snow then sent *The American Moslem* throughout India and
wrote letters to the Muslim leaders condemning the American Islamic Propaganda.[111] Webb later stated, "These people
printed in their papers that I had received that much money,
and so the impression that I really did got fixed in India and
traveled back here. They also poisoned the minds of Abdulla
Arab and his associates, and they broke their contract with
me, leaving me absolutely without funds."[112]

---

[109] Bennett, *Anthony Comstock*, 1023-1024.
[110] "The Call to Prayers," *Middletown Daily Times*, December 11, 1893,
    p. 1.
[111] "Fall of Islam in America," *New York Times*, December 1, 1895, p.
    21.
[112] Ibid.

Webb was convinced that his rivals only wanted to reap financial gain by his demise. In 1895, he wrote: "Since we began our efforts in the United States, nearly two years ago, we have encountered hundreds of people, male and female, who have schemed and lied vigorously in order to gain the confidence of the Oriental Mussulmans and thus reap a profit."[113] Notwithstanding, Webb persevered and continued his work in a modest fashion during the first half of 1894, speaking in Pennsylvania, Washington D.C., and local New York events.[114] A scaled-down publication, *The Voice of Islam*, began publication in June from the new headquarters at 30 East 23rd Street in New York.[115] By May 1894, for lack of receiving his contracted salary, Webb was hurting personally and could no longer afford to keep his family in New York City. His wife was able to use her personal funds to purchase a rock-strewn 10-acre farm in Ulster Park, Ulster County, New York for $1,000. Webb remained part-time in the city at the residence of H. Ali Lewis, business manager of *The Voice of Islam*.[116] Of course, many in the press and in Webb's own organization who had not seen the Ulster Park property believed he had purchased an opulent estate.[117] In

---

[113] "News Notes," *Moslem World and Voice of Islam*, February 1895, p. 4.

[114] Webb spoke at: Williamsport, PA, January 9, 1894. "Advertisement," *Daily Gazette and Bulletin* (Williamsport, PA), January 8, 1894, p. 5; Blavatsky Branch of the Theosophical Society, Washington D.C., March 8, 1894. "Mahommed Webb Defends Islam," *Washington Post*, March 9, 1894, p. 6; Bradford, PA, April 15, 1894. "Brevities," *Evening Democrat* (Warren, PA), April 16, 1894, p. 4; Nineteenth Century Club, New York, April 17, 1894. "Meeting of the Nineteenth century Club," *New York Times*, April 17, 1894, p. 5; and International Temperance Conference, Staten Island, New York, June 4, 1894. "Prohibitionists in Session," *Constitution* (Atlanta), June 4, 1894, p. 2.

[115] "Trouble in Islam," *New York Daily Tribune*, July 14, 1894, p. 12.

[116] Ibid.

[117] "Mohammedan Webb Buys a Retreat," *New York Times*, May 29, 1894, p. 9; and "Fall of Islam in America," Ibid., December 1, 1895, p. 21.

reality, as described by a *New York Times* reporter the lowly home was "...a miserable house without step, stoop, or porch, set in the middle of a rough, uneven patch of ground that was fenced off from the road by a few strands of wire that dragged most wearily...There was no sign of garden or lawn—just a hilly, lumpy, rough yard, that ran back 100 feet or so until it lost itself in a clump of woods at the back."[118] Nonetheless, news of the purchased land coupled with unpaid salaries at the office of *The Voice of Islam* set the stage for a bizarre confrontation.

In mid-July, Nafeesa M.T. Keep, secretary of *The Voice of Islam* locked Webb and all other staff out of the office of the paper. Keep's animosity towards Webb had been building for some time due to her contention that Webb had mismanaged the funds from both India and subscriptions to the paper.[119] She alleged "While Mr. Webb lived in luxury and denied himself nothing...her modest salary had not been paid and the office rent was long overdue."[120] She had apparently begun writing to Turkish officials in an effort to discredit Webb, who was seeking funds from the Sultan's government. She proposed to stay locked in the offices until she heard back from the Turkish officials or Webb resigned as president of the American Moslem Brotherhood and the publications were turned over to her charge.[121] Accusations of theft brought by Webb were countered by Keep's accusations of conspiracy and mail fraud.[122] These charges, brought before the authorities amounted to little and Webb took pos-

---

[118] Ibid.

[119] "Muhammed Webb Locked Out," *New York Times*, July 14, 1894, p. 5.

[120] "Trouble in Islam," *New York Daily Tribune*, July 14, 1894, p. 12.

[121] "Muhammed Webb Locked Out," *New York Times*, July 14, 1894, p. 5; and "Nefeesa Keep Breakfasts," Ibid., July 16, 1894, p. 1.

[122] "She Wants a Warrant for Mohammed Webb," *New York Daily Tribune*, July 27, 1894, p. 12; and "Nefeesa M.T. Keep Retaliates," *New York Times*, July 30, 1894, p. 9. The accusation of mail fraud was for a letter Webb wrote appealing for funds abroad. "Nefeesa Keep Breakfasts," Ibid., July 16, 1894, p. 1. For text of fundraising letter, see Appendix D.

session of the company's entire holdings and removed them to his Ulster Park home.[123]

It had been reported that Webb contemplated a return trip to India, Egypt, and Turkey in October, no doubt to clear his name and raise funds. However, it appears this did not occur for he was in Ohio lecturing in early November.[124] In January 1895, the monthly publication *The Moslem World and Voice of Islam* was first issued from The Moslem World Company's new headquarters in Ulster Park, operated by Webb and his son Russell.[125] Summing up Webb's determination to continue his mission in the first issue he wrote: "Notwithstanding Hajee Abdulla's failure to comply with the terms of his contract, and in spite of the vigorous, persistent and utterly unprincipled efforts of our enemies to destroy our mission, I have been enabled, with God's help, to continue the good work undertaken for Islam, and the results have fully equaled my expectations. God put it into the minds of faithful Mussulmans to come to my relief when help was most needed, and if He spares my life, I will perform, to the best of my ability, all I promised to do under my contract with Hajee Abdulla Arab. God knoweth the hearts of all men and will judge us both."[126] In this theme the journal continued, as did the lectures and parlor talks, and the American Moslem Brotherhood was able to hold its annual elections for officers.[127]

---

[123] "Son of the Prophet at Rondout," Ibid., August 17, 1894, p. 1.

[124] Mention of trip abroad in "Advertisement," *Daily Gazette and Bulletin* (Willimasport, PA), January 8, 1894, p. 5. Ohio lectures in "Mohammed Russell in Toledo," *Salem Daily News* (Salem, Ohio), November 8, 1894, p. 1.

[125] "Fall of Islam in America," *New York Times*, December 1, 1895, p. 21.

[126] "To My Oriental Brothers," *Moslem World and Voice of Islam*, January 1895, p. 4.

[127] In May 1895 the parent arm of the American Moslem Brotherhood held elections in New York. The out going officers were: President— Mohammed Alex R. Webb; Vice-President—C. Omar McCoun; Secretary—C. Omar McCoun; Treasurer—H. Ali Lewis; Librarian—R. Othman White; Advisory Board—E.A. Arnold, H. Fatima Peabody, Khaled D. Hutchins, and Ferhad Ezzet; Board of Publications—

Webb retained many allies in India and across the Muslim world who wrote letters of support and sent contributions, the main financiers being Viqar-ul-Umra, Prime Minister to the Nizam of Hyderabad, Ahmed Moola Dawood of Rangoon, and Kazi Shahabuddin Khan of Poona, all of whom Webb met while in India.[128] The Muslims in Durban, South Africa held a special meeting in reference to financial support for Webb and were assisted by the only non-Muslim in attendance, a young Mohandas Ghandi who helped translate Webb's letter into Gujerati.[129] Webb also retained support in some of the Indian press, with the *Mohammedan Observer* noting: "It is regrettable that, at the threshold of the great work, foes from within—professing Moslems—and without, should set themselves to the task of disseminating erroneous reports to shipwreck the propaganda."[130]

The Moslem World and Voice of Islam continued to be published in its diminished form at a financial loss throughout 1895, quelling plans to separate the journals into their original formats in January 1896. In an effort to promote financial transparency, pleas for funds were accompanied by a request to send money directly to their printer, Edgar S. Werner, instead of the Moslem World Company.[131] Nonetheless, the same charges of financial impropriety arose again in late 1895 when the Nawab of Basoda who had no previous dealings with Webb decided to inquire about the Propaganda for himself while visiting the United States. The

---

Ahmed Hamouda, H. Ali Lewis, R. Othman White, and C. Omar McCoun. The newly elected officers were: President—Mohammed Alex R. Webb; Vice-President—R. Othman White H; Secretary—Ali Lewis; Treasurer—C. Omar McCoun; Librarian—Ahmed Hamouda; Board of Directors—Mohammed Alex R. Webb, H. Ali Lewis, R. Othman White, C. Omar McCoun, Ahmed Hamouda, Ahmed Hamouda, H. Fatima Peabody, and August Berg. "Annual Election," Ibid., April 1895, p. 4; and "Annual Election," Ibid., June 1895, p. 3.
[128] "Mohammed Webb's Account," *New York Times*, March 27, 1896, p. 3.
[129] "Earnest Moslems," *Moslem World and Voice of Islam*, October 1895, p. 3.
[130] "News Notes," Ibid., March 1895, p. 3.
[131] "Special Notice," Ibid., October 1895, p. 2.

Nawab spoke with Webb's nemesis, Lant, but not Webb during his half-hearted investigations, which resolved little, and Webb was exonerated in the end. An exposé into the matter published in the *New York Times* answered many questions and proved that Webb was barely scraping by in Ulster Park not living in the luxury previously described.[132] Furthermore, the enormous sums of money that Webb allegedly received were so out of line that Hajee Abdulla Arab sent a letter to the editor of the *New York Times* listing the number of rupees he sent. Webb responded with an exact accounting of major donations received from India listed in dollars, a total of $11,427.51 over a nearly two-and-a-half year period. Arab's letter to the editor seemed to hearten Webb who then conciliatorily referred to Hajee Abdulla as his "...good brother..."[133]

But the damage had been done to their relationship and to the American Islamic Propaganda. Hasan Ali later reflected on the demise of the mission stating, "It happened just as I had said. The Muslims of India did promise donations, but the collections were few. Haji Abdullah Arab tried his utmost but with little success. The Muslims were ready to spend millions on un-Islamic pursuits but gave nothing for this important task. The only worthwhile collections were made in Rangoon, and Hyderabad Deccan. I think the total amount that was sent to Mr. Webb was rupees thirty thousand, out of which Haji Abdullah Arab had contributed rupees sixteen thousand. Poor Haji was crushed in this undertaking."[134] Some of the details concerning the fallout between Hajee Abdulla Arab and Webb are uncertain, but it is clear that the demise of the mission coincided directly with Arab's withholding of funds.

Although the financial problems were major, Webb also contributed to the collapse of the Islamic mission. His main failure was the inability to truly imbibe the teachings of

---

[132] "Fall of Islam in America," *New York Times*, December 1, 1895, p. 21.

[133] "Webb's Supply of Rupees," Ibid., March 20, 1896, p. 4; and "Mohammed Webb's Account," Ibid., March 27, 1896, p. 3.

[134] Ahmad, *Mujaddid Azam*, 160.

46

Islam with regard to class and race—all men are equal, separated only by their degree of faith. Instilled in him were the inherent class and racial biases of a late nineteenth-century American white man of privilege. His insistence that Islam could only be understood by the educated elite left untapped the masses of lower class whites, immigrants, and most importantly African-Americans. The African-American community was fertile ground for conversions, as evidenced by the Islamic movements of the twentieth century.

When asked if he had sought converts near his rural Ulster County home, his reply was: "No, you can only hope for converts among thinking people, to begin with. The people here are not educated to think."[135] With regard to immigrant Muslims living in New York it was said, "There are a few Mussulmans in New York, but they are mostly peddlers and low-caste Hindus, and Muhammed Webb will not associate with them." He was later quoted in the same article as being sure the poor condition of Muslims in the east was due to many factors including "...racial and climatic influences."[136] Furthermore, his writings and travel journals are replete with disdain for lower class Muslims. It is clear that by choice and ideology Webb left himself only a small window for success, pinning his hopes on a meager cohort of the American intelligentsia.

Life After the Mission (1898-1916)

In 1898, Webb and his family moved to Rutherford, Bergen County, New Jersey, and settled in at 391 Orient Way. The records concerning the last eighteen years of his life are murky, but a few details are known. Webb reestablished himself as a newspaperman by purchasing the *Rutherford News* in 1898, but this venture was short-lived and he sold the paper to a larger news company in 1900.[137] In Rutherford, Webb became civic minded and politically active as

---

[135] "Fall of Islam in America," *New York Times*, December 1, 1895, p. 21.

[136] "Muhammed Webb's Mission," Ibid., February 25, 1893, p. 1.

[137] Haberly, *Newspapers and Newspaper Men*, 10-11.

he had been during his years in Unionville, Missouri. At various times Webb served on the Rutherford Board of Education, acted as foreman of the Bergen County Grand Jury, and as president of the Democratic Society among other positions.[138] He also became an active member of the Rutherford Lodge of the Knights of Pythias, a non-denominational fraternal order. He served as chancellor of the Rutherford Branch as well as on the statewide Finance Committee.[139]

Webb remained an advocate of Islam and especially the Sultan of Turkey for many years after his formal mission had become defunct. He made very few public appearances in this role between 1898 and 1901. In 1901, the Sultan of Turkey, Abdul Hamid II, appointed Webb as Honorary Consul General of the Turkish government at New York and when he visited Turkey he was given the decoration of the third Order of Medjidie and the Medal of Merit, as well as the title Bey. His appointment letter stated, "...not only for the enumeration of your services, but for your strong efforts to build a mosque in America, as well as a cemetery for the benefit of Islam, to which you have converted many Americans."[140]

The main impetus behind these honors went beyond mere rewards for Webb's attempt to establish Islamic intuitions in America, they were a reward for his continued loyalty and defense of Turkey and the Sultan. Webb had long been an avid supporter of the Sultan who also claimed the title of Khalifa, or leader of the Muslim world. In 1893 the news sections in the *Moslem World* often spoke in defense of Turkey and Abdul Hamid II.[141] Webb took a particularly strong stance against criticism of the Sultan with regard to

---

[138] Ives, "Story of the First American Convert to Islam," 7.

[139] "Obituary—Alexander R. Webb." *Rutherford Republican*, October 7, 1916, p. 1. Finance Committee information came from "Committees for Pythians," *Trenton Times*, February 20, 1903, p. 7.

[140] "Sultan Honors an American," *New York Times*, October 1, 1901, p. 9; and quote comes Ives, "Story of the First American Convert to Islam," 7.

[141] A sampling of articles include: "A Modern Hero," *Moslem World and Voice of Islam*, September 1895, p. 3; and "Gross Inconsistency," Ibid., December 1895, p. 1.

accusations of atrocities against Armenians and other Christians in Ottoman lands. 1895 he wrote an extensive letter to the *New York Daily Tribune* arguing against the public perception of the Sultan as a "...religious despot."[142] Webb was the presumed author of a thirty-five-page pamphlet entitled, *The Armenian Troubles and Where the Responsibility Lies*, published under the pseudonym, "A Correspondent", as well as a 67-page work entitled, *A Few Facts About Turkey Under the Reign of Abdul Hamid II* by an "American Observer." Both of these works were promoted and made available for purchase by the *Moslem World and Voice of Islam*.[143] As early as 1898, Webb was acting as an agent for the Turkish government, approaching Richard Gottheil, Chairman of the Federation of American Zionists, on behalf of the Sultan to dissuade Gottheil from his Zionist efforts in Palestine.[144]

Webb's new consul position reenergized his dormant propaganda efforts once again. Soon after returning from Turkey, dressed in full Turkish regalia, he gave a lecture in Rutherford on Constantinople and Turkish home life.[145] In succeeding years he also gave lectures on Islam at the Theosophical Society of New York despite having formally resigned from the group in 1897.[146] In 1907, Webb gave a lecture on Thomas Paine entitled, "Paine as Seen by an American Adherent of Islam," to The Sunrise Club of New York.[147] In 1910, he assisted in revising the posthumous

---

[142] "In Defence of the Sultan," *New York Daily Tribune*, December 23, 1895, p. 5.

[143] A correspondent, *The Armenian Troubles and Where the Responsibility Lies*, New York, 1895. A note in the periodical *The Open Court* (Vol. 10, 1896, p. 4998) has Webb listed as the author of the work; and "News Notes," *Moslem World and Voice of Islam*, May 1895, p. 3.

[144] Moshe Davis, ed., *With Eyes Toward Zion*, 267.

[145] Haberly, *Newspapers and Newspaper Men*, 11.

[146] "Classifieds," *New York Times*, February 28, 1903, p. 6; "Classifieds," Ibid., October 29, 1904, p. 11; and membership information from David Wietersten, Theosophical Society Archivist, e-mail message to editor, February 16, 2004.

[147] "No Brand of Anarchy Proves Too Startling for The Sunrise Club Twelve-Bits Dinner," *Washington Post*, February 10, 1907, p. 10.

translation of Mirza Ghulam Ahmad's 1910 publication, *The Teachings of Islam.*[148]

In Webb's last years he was head of the sales department at Martindale Mercantile Agency in New York City. He had long suffered from diabetes and after returning home from work one Saturday afternoon he complained of illness and died the following morning on October 1, 1916 at the age of sixty-nine. He had been in association with the Unitarian Church at Rutherford through his wife who had turned away from Islam and it was the Reverend Elizabeth Padgham of this church that presided at his funeral. The pallbearers were members of his lodge of Pythinans. Webb was buried at the non-sectarian Hillside Cemetery.[149] His daughter, Mary, later confirmed that despite the funeral arrangements Webb had indeed died believing in the truth of Islam.[150] Ella G. Webb lived another three-and-a-half years passing away on April 10, 1920 and was buried next to her husband at Hillside.[151]

Brief Overview of Islam in America After Webb's Mission

During the two succeeding decades after Webb's Propaganda movement ended, the stream of Arab and other Muslim immigrants increased. However, it is was not until the aftermath of World War I and the collapse of the Ottoman Empire that Muslim immigration began to become statistically significant. Turks, Kurds, Syrians, Albanians, Russian and Yugoslav Muslims, and others immigrated and formed small ethnic communities in the mostly urban areas of the Northeast and Midwest. The number of Muslim immigrants, although greater than before, was stunted due to rigid quota laws required by the Immigration Regional Restriction Act of 1917. Specifically, the Asiatic Barred Zone provision

---

[148] Ghulam Ahmad, *The Teachings of Islam*, ix.
[149] "Obituary—Alexander R. Webb," *Rutherford Republican*, October 7, 1916, p. 1.
[150] Ives, "Story of the First American Convert to Islam," 8.
[151] "Obituary Notes," *New York Times*, April 11, 1920, p. 20.

excluded South Asians, Central Asians, and Southeast Asians from the United States, leaving mainly European Muslims and Arabs within the British and French Mandate territories to immigrate.

Once in America, the immigrants did little to spread their religion to the larger society. Instead, they formed insular ethnic and religious institutions. The spread of Islam amongst native-born Americans did not take hold until the 1920s and 1930s, as several quasi-Islamic movements such as the Moorish Science Temple and Nation of Islam took root in the expanding urban African-American communities of the North. These groups deviated greatly from orthodox Muslim teachings and practices, but some of their terminology and beliefs were recognizable as Muslim. As well, during this period the Indian-based Ahmadiyya Movement in Islam conducted missionary efforts among African-Americans and made gains, but much smaller than the two other movements. Egyptian-Sudanese Muslim Duse Muhammad Ali also worked hand in hand with Marcus Garvey's UNIA movement and used Islamic identity and imagery to further Black Nationalist goals. Although these groups did not have a lasting institutional impact on their contemporary communities, they laid the groundwork for future generations of African-Americans to convert to orthodox Islam. For instance, in 1937 Wali Akram, who originally joined the Ahmadiyya Movement but later left the group, helped form the First Cleveland Mosque, the first orthodox mosque established solely by native-born Americans.[152]

During the middle decades of the twentieth century, the aforementioned trend of trickling Muslim immigration and expansion of Black Nationalist Islamic movements continued. A sea change in American Muslim history occurred during the late 1950s and early 1960s when Malcolm X brought the Nation of Islam out of the shadows, offering an alternative to the vision of the established Civil Rights leaders. Due to Malcolm's influence, the Nation of Islam expanded exponentially. Malcolm X converted to orthodox Islam prior to his assassination in 1965. Similarly, most of the

---

[152] Dannin, *Black Pilgrimage to Islam*, 87-115.

followers he brought to the Nation of Islam eventually converted to orthodox Islam in the mid-1970s after the death of the group's leader, Elijah Muhammad.

Even more significant to the growth and diversification of Islam in America was the liberalization of immigration laws in the mid-1960s. The new laws allowed Muslims from the Middle East and South Asia to immigrate in unprecedented numbers, without exclusions. These events four decades earlier formed the foundation on which contemporary American Muslim institutions and communities were built. Now Islam is a growing religion in the United States, something Webb envisioned a century earlier, but was never able to bring to fruition.

Conclusion

Webb's Islamic mission to America came between two distinct periods in American Muslim history, the antebellum era of enslaved African Muslims and the rise of quasi-Islamic groups in the early twentieth century. Although he left no lasting movement or institutions, and no link between the other periods of Muslim American history, his significance as the first purveyor of Islam to the larger society outweighs his failures. Before Webb, Islam in America was all but unknown in the general public. But during his mission, nearly every paper in America covered his activities and beliefs to one extent or another, and in New York, he generated a great deal of attention.

Webb wanted nothing more than to break down the prejudices of the American people towards Islam, and in this regard he partially succeeded. He spoke in favor of Islam at a period when nearly all public comment upon the faith was uninformed claptrap. He was a media-savvy white man of position, which allowed his voice to be heard when lower-class whites, African-Americans, and immigrants could not. Many critics believed that Webb had set out on a mission that had no possibility of success; however, it can be argued that his failure was due largely to his own personal idiosyncrasies, biases, and goals, and not simply because of Islam. Had Webb selected an appropriate method, message, and audience for his mission, he might have bought time to establish institutional foundations for the religion. As was the case with the Nation of Islam, which simmered for decades unnoticed until it boiled over under the right conditions in the late 1950s. Unfortunately, Webb's narrow focus on white intelligentsia and failure to manage the American Islamic Propaganda effectively doomed his cause more than any perceived inherent incompatibility between Islam and America values.

Webb and his efforts to promote Islam in the United States were all but forgotten for more than four decades. In 1943, with the assistance of his daughter, Mary, a group of Muslims in New York held a commemorative meeting and lecture on what would have been Webb's ninety-sixth birthday.[153] The Muslims in question had no relation to Webb or his earlier propaganda work and were just as surprised to discover his story as were modern scholars who have rerediscovered him. In 1945, Dr. Emory H. Tunison, also involved with the commemoration, wrote an article about Webb for *The Arab World*.[154] Until the past decade, no scholarly attempt to analyze Webb's life or mission has been written. He is only now being mentioned in the historical record of Muslims in the United States.

One major problem with research concerning Webb and his contemporaries is the lack of primary materials. Because Webb and his associates corresponded with Muslims from all over the world, many of these documents are in archives and private collections outside the United States. Fortunately, as interest in the early American Islamic movements expands, more historians are dedicating their energies to unearthing the documents from this period. In time, our knowledge of late nineteenth century American Muslims will

---

[153] The text of the lecture given by Nadirah F. Ives Osman was published in: Ives, "Story of the First American Convert to Islam."

[154] Emory H. Tunison, "Mohammed Alexander Russell Webb, First American Muslim." *The Arab World* 1 (1945): 13-18. This article is basically a summary of Ives' article mixed with quotes from Webb's speech at the *Parliament of Religions*.

be clearer, as will the other chapters in the Islamic presence in the United States.

As far as Webb is concerned, his future stature may be raised or lowered as more information on his contribution and the contributions of others comes to light. Like the enslaved African Muslims, Webb was unable to create long-standing Islamic institutions in America. Nonetheless, as with the enslaved Muslims, Webb's presence is of historical significance, and he has solidified his place as a pioneer in American Muslim history.

I.

MANILA

August 29, 1892

Went to the hotel of Lala Ary in the Escolta of Manila.[155] Sent the last of my furniture to Genato's to be sold at auction.[156] Have divorced myself from the Consulate. Lala's is a queer place. Chinese shoemaker's incessantly at work in the Saguan. Received 5 letters from my darling wife yesterday. She's a good woman. God bless her. Met George Simpson at Lala's. An old man of 65—an Englishman and a Theosophist. Ate meat to-day at tiffin at Mrs. Mustard's for the first time since 1882. Ate it again at dinner at Lala's. Dr. Donelan advised it as I had been ill for several weeks with bowel complaint. Excessive diarrhoea every morning. It is singular that I have been more or less ill since the 8th of June last, the day on which my wife and precious babies left Manila for the States.[157] I think that my wife's presence is necessary to me. I am certainly very lonely without her. I hope to be bet-

---

[155] Fonda de Lala was located at No. 37 Barraca St. in the Escolta. José Rizal, *Noli Me Tangere*, 434.

[156] Manuel A. Genato was the proprietor of an auction house in the Escolta. *Commercial Directory of Manila*, XXXV.

[157] Webb's family was sent to a fruit ranch 40 miles from San Francisco to await his arrival in New York. Webb, *Lectures on Islam*, 17. For the text of the letter concerning this matter, see Appendix E.

ter to-morrow. The dinner was very good and I think that my
health will improve with this kind of board. Had a restless
night, the Chinese shoemakers talking till after midnight.
Very good hotel—Lala speaks English, French, Spanish &c.

August 30, 1892

Put in a busy day at the Consulate. Young Ordonez,
Daland's clerk commenced.[158] Wrote to Ella, Bessie and
Judge Knott.[159] Dined for the 1st time in the evening at 6
with Mr. Simpson. I have never tried to keep a journal before
and am too lazy, I guess to make a success of it. There's
nothing in the world to write about. Took ice-cream at 7 p.m
at Restaurant de Paris with Bournes and Worcester.[160] Very
good cream. Returned to Lala's at 8:15 and had a chat with

---

[158] William A. Daland, 1856-?, native of Brooklyn, N.Y., arrived in Manila in 1880 and was engaged in business as a broker. He was appointed interim American Consul upon Webb's resignation on Sept.
5, 1892. *Despatches fromU.S. Consuls in Manila,* Reel 10, September
5, 1892; and "List of American Citizens."

[159] Ella was Webb's wife and Bessie his step-daughter. Judge E.W. Knott
was one of Webb's old friends from St. Louis. He worked as a deputy
insurance inspector for the state of Missouri, and was the brother of J.
Proctor Knott, Governor of Kentucky. In an article Webb referred to
Knott as "…one of the most admirable and loveable men I ever met.
A genial, whole souled person whose principle motive in life apparently, was to shed sunshine upon the lives of those around him and to
make those who enjoyed his unselfish friendship better and happier
for having known him." Alexander Russell Webb, "Two Remarkable
Phenomena," 248; Letter from F.M. Cockrell to Thomas F. Bayard,
23 September 1887, "Letters of Application and Recommendation.";
and Letter from D.R. Francis to Thomas F. Bayard, 17 September
1887, Ibid.

[160] Dean C. Worcester, 1866-1924, and Dr. Frank S. Bourns, zoological
researchers from the University of Michigan who in 1892 were in the
later part of the 2 ½ year Louis Menage expedition. This was their
second trip to the Philippines and both would later play roles in the
American colonization of the islands beginning in 1898. Dean C.
Worcester, *The Philippines Past and Present,* 1-7; and Christopher
Winters, ed., *International Dictionary of Anthropologists,* 768.

Simpson. He's a queer genius—keeps his feet up to ease his heart. Says that people with cold feet never think of the effect on the heart. Went to bed at 9.

August 31, 1892

Went to the Consulate this morning and put in the forenoon on my final report. In the afternoon the Consulate was removed to the Hong Kong and Shanghai Bank building to good light rooms. In the evening called at Glazebrook's and came home at 9:45.[161] The daughter of the dressmaker opposite the hotel was celebrating her saint's day and a dance was in progress making sleep impossible so I had a pleasant chat with Simpson. Eccentric old genius. He told me a story of one of his boy friends who was converted by having a bent horseshoe nail at the end of a kite string hooked in his coattail as he started for home over a lonely road. Thought the devil had got him and he was converted and became a preacher.

September 1, 1892

Spent the greater part of the day at the new Consulate arranging and putting in order the books, blanks and furniture with the assistance of Ignacio and Nicholas. About 5:15 went to Paterno's to get the rent receipt straightened out and from there went to hotel. After dinner called on Mr. and Mrs. Knudsen, and had a pleasant chat of about half an hour with them.[162] The immense mango tree near their house has been cut down to make room for another house. Went to lunetta

---

[161] Probably Frederick Edward de Tweenbrook Glazebrook, 1859-?, a native of Lancashire, England. He arrived in the Philippines in 1885 and was employed as a clerk with F.M. Heriot and then with Findlay, Richardson & Co. from 1887-1893. *South-East Asian Biographical Archive*, fiche 128, frame 230-31; and J.J. Howard and F.A. Crisp, eds., *Visitation of England and Wales*, 3:59.

[162] Frederick Knudsen, 1830-?, and Catherine Knudsen, 1842-?, Americans who came to Manila in 1887. "List of American Citizens."

afterward and heard the band play two pieces.[163] Returned to Escolta and attended the auction sale of some of my goods at Genato's. Price's very low, Mrs. Vidals bed sold for $11.50. Returned to hotel at 10. A dreadfully uneventful day.

September 2, 1892

Went to the Consulate at 7:15 and wrote application for passport, letter to Dept. of State and other office work. At 10, went to the Government office in Malacañan and secured my passport. Took a siesta and put in the greater part of the afternoon in office work. In the forenoon at 11:30 went to the German Consul's and had him certify to my verification of Treasury fees. In the evening accompanied old man Simpson on a call on Bridger and Hayden.[164] Had a pleasant chat until 9:10 and returned to hotel. All this reads very much like a school girl's diary but I feel so little like writing and there is so very little to write about that I can hardly make it much better. At all events it is not intended for anyone's eyes but mine and as a sort of note-book for future writing. The Chinamen in the Saguan are pegging away industriously at their shoes and seem to take life as a very serious thing. Perhaps it is—who knows?

September 3, 1892

Really too busy to write. Went up to jail to see two pirates arrested at Ponape.[165] Consul at Hong Kong thought they

---

[163] La Lunetta as described by Worcester is "… is a drive and promenade, the carriage road enclosing an oval piece of slightly raised ground, on which are a couple of band-stands, and numerous chairs and benches for the convenience of the public." Dean C. Worcester, *The Philippine Islands and Their People*, 42.

[164] Likely Charles Bridger who was later Webb's agent in Manila for his publication *The Moslem World*. "Agents for *The Moselm World*," *Moslem World*, August 1893, p. 8. Hayden is unidentified beyond Webb's account.

[165] Ponape was located in the east Caroline Islands.

might have stolen the missing schooner "O.S. Fowler".[166] Received returns from auction $87.75. At night dined at Glazebrook's with Williamson—afterward Fowler came in. Left at 11. Too late.

September 4, 1892

Put in the morning closing up my reports at Consulate and the afternoon in repacking trunks. Late in the afternoon an American mail arrived with letter from Ella. This is my last day as Consul. To-morrow I will be a plain American citizen. In the evening paid my farewell call at Mustard's and had a pleasant chat until 9 when they walked down with me nearly to Santa Cruz. I returned to hotel wishing that I was going straight to San Francisco instead of to Singapore. Well, God's will be done.

September 5

Spent the greater part of the morning in getting my ticket and money arrangements fixed for to-morrow. Mailed letter to Ella containing two drafts and wrote another. In the afternoon read "Heralds" at the Consulate until nearly 4 then called on the Glazebrooks and said good-bye. Also bid good-bye to Blodgett, Barnes, the two Barrettos, Capt. Cobban, and Bruce.[167] About 4:30 Mrs. Sinclair and Mary called at the hotel with a package of books which they wanted me to

---

[166] Consul at Hong Kong was most likely Oliver H. Simons, a native of Colorado who was appointed to the position in 1889. Susan Holly, U.S. Department of State, Office of the Historian, e-mail message to editor, July 20, 2004. The O.S. Fowler was a schooner out of San Francisco named after the noted phrenologist.

[167] Blodgett probably refers to E.W. Blodgett, 1857-?, born in New York and arrived in Manila around 1881. "List of American Citizens." The two Barrettos possibly refers to R.E. and A.M. Barretto, of the Ecuadorian consulate. *Commerical Directory of Manila*, 31. Cobban was captain of the *Zafiro* a steamer for the China & Manila Steamship Co., which made runs from Hong Kong to Manila. John J. Aubertin, *Wanderings & Wonderings*, 307. Others mentioned were not identified beyond Webb's account.

take to Singapore for Capt. Darke.[168] During the afternoon the barometer fell rapidly and at 6 p.m. the wind was blowing a gale and the rain fell in torrents. A typhoon threatens.

September 6, 1892

The wind blew rather stiff all night and sighed and moaned threateningly among the trees outside my window. About 3 o'clock I arose and looked out; the clouds had broken a little but the wind still blew in strong fitful gusts. At 4:30 I arose again, went to the closest and then took my bath. By daylight the wind had fallen off a little and the clouds were breaking away. I saw that it would be possible for the steamer to sail and that I would at last, God willing, get out of Manila. My stomach had almost entirely recovered its normal tone. At 7 I started for the steamer which was lying in the river off the Magallanes monument. Mustard came up in the "Holdfast" came aboard and chatted for half an hour. I was not favorably impressed by the Capt. nor the steamer. The former seemed lazy, listless, incompetent and disagreeable generally. We were to have started at 9 but mail failed to arrive. Last view of Puente de España, Paseo de Magallanes, the old walled city, Macleod's Smith, Bell and Co, Warner Blodgett & Co, Ker and Co's go-downs.[169] At 9:15 mail arrived and whistle blew to start. At 9:25 began to swing round into stream. Old Jim Allison on the "Britnan," past the old fortified wall, the paseo of Malecon and monument to Simon de Anda. The port works and fort at end of sea wall. Pilot left us at 10 and we started fairly on our voyage leaving Manila in the distance. Last glimpses of La Lunetta, the Santa Yglesia Cathedral, Jesuits church, tall spires of San Sebastian. Masses of drift-wood in the Bay showing that there had been rough weather outside. It was not without a slight feeling of regret and sadness that I saw Manila fading away in the dis-

---

[168] Captain F.M. Darke, Marine Superintendent of the Straits Steamship Company, Singapore. K.G. Tregonning, *Home Port Singapore*, 34.

[169] Several companies with these or similar names existed in Manila: Macleod & Co.; Warner, Barnes & Co.; Smith, Bell & Co.; and Ker & Co. *Commerical Directory of Manila*, 106 and 118.

tance forever. I had spent many happy hours there as well as unhappy ones. It had been my home for 5 years and a good home too. Now all was broken up and I was about to begin the world anew. The "Kowshing" started for Hong Kong about the same time and we steamed along side by side until past Corregidor.[170] Just before 11 it began to rain and the heavy clouds shut down over Manila cutting the city off from view entirely. Curious rocks off Corregidor passing to the South. "Kowshing" took north passage and left us. Soon after passing Corregidor wind increased to a gale and rain fell in torrents. I soon began to feel "that tired feeling" and had to lie down on the companionway benches. There I staid till dark, most of the time and then started for my room. We had a cargo of copra on board the odor of which added to the usual nastiness of a Spanish steamer made the cabin unbearable. Took off my coat and vest and bunked in on the companionway benches where I got some fresh air. Passed a miserable night. The steamer pitched and tossed violently and the loose articles in the cabins and saloons pitched and crashed about at a great way. It was the roughest weather I ever experienced at sea.

September 7, 1892

I went down stairs about six aroused the sleepy mayordormo, and told him I wanted a bath. "Sea or sweet" he asked. I thought it would be easier to get the sea water than the fresh so I told him to get that. After about ¾ of an hour he managed to get four pailsful into the dirty tub and I got a sort of a kind of a bath but it made me retch as if I was going to vomit. Managed to eat desayuno and lay in my berth until 10 when breakfast was announced. I couldn't eat. Sea still rough and no signs of clear weather. 3 dogs, two cats and litter of kittens, 2 goats 2 pigs and 1 cow on board. All but the cow

---

[170] The *Kowshing* was owned by the British firm Jardine, Matheson and Co. It was sunk by the Japanese on July 25, 1895 while transporting 1,100 Chinese troops in one of the first acts of aggression during the Sino-Japanese War of 1894-1895. S.C.M. Paine, *The Sino-Japanese War of 1894-1895*, 132-4.

roam around at will and mess the deck up. But the officers don't mind it; they have evidently become indifferent to dirt. The Dr. eats birds bones and all. Remained on my back most of the day on the companionway benches. 1st mate said surely that the sea would be calm by noon as we would then be out of the typhoon circle. He was mistaken. All day and all night it blew great gusts and rained in torrents at intervals. Poor boys, some of them sea sick had to keep at work running up and down stairs with drinks for the officers and engineers. They seemed to want to drink every 15 minutes. Poor natives treated like animals with little or no rest. At night I had my bed made in the companionway and it was very comfortable and cool. Plenty of fresh air and no bad smells. As I had slept a good deal during the day my sleep was much broken all night by the violent motion of the vessel and the noises of various kinds that abounded about the companion-way.

September 8, 1892

Arose at 6 after some trouble managed to get a sort of sponge bath and lay down in my berth for awhile with one of the port holes open. But had to close it on account of the rain and high seas. Had desayuno at 7 and got up into my seat in the companionway as soon as possible. 1st mate came up and found one of the cabin boys lying on deck near the main mast, sea-sick. He aroused him with a kick and a cuff and turned him over to the steward who treated him similarly. Sea still very rough and wind strong with no signs of abating. Passed the islands of Paragua and Puerta Princesa steaming slowly.[171] Strong head winds. Shortly before 10 the sun came out partially and the wind fell off a little. About 1 the sea had quieted down considerably and it was quite comfortable sailing. The cow was killed this morning to supply the table with beef. Poor thing she didn't seem to enjoy the trip and death was probably a happy release for her. The food gets worse and worse. Very little more now than beef

---

[171] Should be "Puerto Princesa," which is not an island, but rather a key port city on the Island of Paragua.

and pork cooked in various ways. Spent most of the day on deck enjoying the cool breeze. Only one or two light rain squalls. After about 3 hours pleasant weather shortly after 2 it began to rain and blow again and we were driven under cover. One of the dog's was sea-sick the first day out and it was pitiful to see him staggering about the deck vomiting like a human being. He looked up at me mournfully as if to say: "This is hard, ain't it?" The Dr. does not take a very deep interest in passing objects. I asked him what a certain island was called "No se" (don't know) was his reply. He has been traveling this route for two or three years twice a month and yet apparently does not know the name of a single one of the numerous islands we pass in the China and Sulu or Mindoro seas. The second day out I asked him how far we were from Manila. "No se," came as usual. And yet he seems to be an ordinarily intelligent man and talks pleasantly readily and fluent on commonplace topics. The day closed calm and pleasant and I slept in my cabin for the first time with the port holes open.

September 9, 1892

Fourth day out from Manila and according to the Capt. we are not half way to Singapore. But the Capt. and 1st mate seem to be very little better informed than the Dr. Yesterday I asked the Capt. how far we were from Manila and he said about 400 miles. About an hour afterward I asked the 1st mate the same question and he thought we had come about 200. We had then been sailing about 12 hours opposite the islands of Paraguas, which I think are something like 500 miles from Manila S.E. The ship gets nastier every day and will be ready to breed a pestilence by the time we reach Singapore. Pigs dogs and rats "botchri" everywhere and the men walk over it without taking the least notice of it. There is a homelike familiarity too about the native servants on board that I don't like. The mayor-dormo, keeps his clothes his cleaning rags,—which he doesn't use—and all his personal effects in my cabin but hasn't offered to sleep there yet. This morning I found him washing his hands in my wash-bowl. I protested and called the attention of the Spanish steward to the fact; but he only gently reprimanded the fellow and didn't seem to think any harm had been done. Yesterday native sailors deftly made rope out of raw hemp as well as if made by machine. Steward put out line astern for big fish but had John the Baptist's luck. About 9 a m passed the little island of Comidan on the left and the big Balabac on the right and out into the China Sea.[172] Beautiful Balabac, grand scenery, indented by coves and bays and mountains wooded to the tops. Sylvan paradise.

September 10, 1892

All night the wind blew and the rain fell in torrents at times. The ship pitched and rolled at a fearful rate and is still doing so with no indications of a break in the weather. How difficult it is to get information about a voyage! When I left Manila one friend predicted positively that in three days we'd be out of the typhoon belt and would have beautiful weather. Another said we'd do it in 24 hours. The 1st mate made it 48 hours. Here it is the 5th day out and we have had bad weather almost continually. At this rate we will be 3 days late in Singapore if we get through at all. In the afternoon the sea quieted down a little and I was able to sleep in my cabin with the port holes open.

September 11

The day opened cloudy with the ship still rolling uncomfortably. The provisions have gone bad and the fare is very hard. I shall be very thankful when I get out of reach of Spanish feed and all things Spanish. I have had more than enough. The sleepy boy—always asleep and someone always in the act of hauling him out of some corner, or off the companion-way house. He has his mat and pillows and sleeps at night with great energy—but still he wants to sleep all day. Toward noon the sea died down to a very comfortable calm and we sailed along most pleasantly. Everything has brightened up and all feel better.

---

[172] Island should be "Comiran."

September 12, 1892

A beautiful day. Rather too hot last night to sleep well and I
changed around from cabin to companion-way house. Managed to sleep fairly well and feel first-rate this a.m. About 9
a.m. we came in sight of the South Natunas Islands which
the mate said belonged to Holland and were used as a coaling station by the Dutch. The Capt. said that all belonged to
England except one small one which belonged to Holland.
The steward said they all belonged to France. The Capt. and
mate discussed the matter at tiffin and the Capt. being Capt.
carried his point, the mate acknowledging that they belonged
to England except the small one used by Holland as a coaling
station. The islands are supposed to be inhabited by savage
Malays. About 2 p.m. passed the North Natunas but only saw
them in the dim distance. The officers think that some time
to-morrow night we ought to reach Singapore but they do not
seem to feel sure of it.

September 13, 1892

Weather clear and bright and the sea comparatively smooth.
The engineer has discovered that he has barely coal enough
to take us to Singapore. Another specimen of Spanish management. The vessel starts to sea with coal enough for a 7
days voyage, the regular running time between Manila and
Singapore and no provision is made for accidents or bad
weather. If we run out of coal we shall be blown back to sea
as the wind is against us and the sails are scanty. It is not a
bright prospect but we'll hope that the coal will hold out.
How such people manage to go to sea and get ashore again is
a problem to me. Bad and shiftless management is the rule.
Nothing for desayuno this morning but tea and dried bread.
They don't toast the bread but put it in the oven and dry it.
Eggs, bananas, mangos, oranges, all gone. There is a water
carraffe in my room but there has been never a drop of water
in it since we started. Night before last my slop can had not
been emptied and I put it out in the saloon where the boy
could smell it. He discovered it in some way, but I'll bet be
never smelled it. He must have fallen over it. He's used to

bad smells and rather looks upon them as necessary append-ages to a ship, like the sails and hawsers. But I'll stop com-plaining. It is such a beautiful day that I should be thankful to God for the many blessings we have rather than complain of the little discomforts. About 5:30 we sighted the Mountain of Pintan and also a steamer bound out from Singapore to Borneo.[173] On the right appeared the Spanish S.S. Santo Domingo en route from Spain to Manila. Of course there was much interest manifested on board our ship and both vessels dipped their flags. In a few minutes we sighted Singapore and felt that our voyage was ended. We passed the light at the entrance to the Strait about dusk and when darkness came on we were steaming up the Strait toward Singapore. I retired at 8 and slept until 11:30 when I awoke and found that we had just dropped anchor to wait for daylight before docking.

---

[173] Should be Bintan.

II.

SINGAPORE

September 14, 1892

I was up long before daylight and after bathing and changing my clothes went on deck to find myself in a more oriental city than I have yet been. On the dock were Chinese, Hindus, Maylays and almost all Eastern types. After getting baggage on dock I tried to bargain with the only "garry" there and he wanted $1.00 to take me up town—four times the regular rate. The fellow who was apparently a Singalese or Malay, seemed determined to rob me so I offered another Malay 20¢ to get me another garry. He did so, and off we started with a bullock cart behind us containing my trunks. For two mortal hours we paraded through the tangled streets looking for 191 Cecil Street and Hajee Abdulla.[174] Everyone we asked seemed to know just who and what we wanted and where we ought to go, but when we got there it always proved the

---

[174] Hajee Abdulla Arab, a Memon merchant who began doing business in Calcutta before relocating to Medina and then Jeddah, Arabia. While doing business in Bombay he heard about Webb's interest in Islam and arranged to meet him in Manila in March 1892 to begin planning the Islamic mission to America. Of Arab, Webb wrote: "He is the grandest, noblest specimen of manhood I have ever met, and a living illustration of the fact that it is possible for a man to be rich, and active in business, and yet be as pure in thought as a child, and as devout a worshipper of the one true God as it is possible to imagine." Alexander Russell Webb, *Islam in America*, 68; and Ahmad, *Mujaddid Azam*, 158-60.

wrong place. The right one was a little further on. Finally an aged Mohammedan told me that Hajee Abdulla was not in town and that I had better go to Raffles Hotel.[175] I did so and was glad of it for it was a very inviting place. There were several detached buildings situated in beautiful grounds laid out in flower-beds, lawns and shrubbery facing the sea or strait. In fact there was nothing to interfere with the view of the water and shipping. A broad macadamized drive passed in front of the hotel up and down which the garries, jinrikshas, and bullock-carts were passing constantly. The view reminded me very much of that from the Grand Hotel at Yokohama but the hotel itself was far more attractively located and, thus far, my impression is that it is a much better kept hotel. This morning I went to the American Consulate and saw Consul Rouncevelle Wildman and delivered Mrs. Sinclair's package at Mansfield and Co.'s for Capt. Darke.[176] The streets are full of interest for me and I expect to find a good deal to write about. My first impression of Singapore is that it is a most beautiful and comely city. It is a vast improvement on Manila. G. Andreas, of G. Andreas and Co., called on me at the hotel at 12 accompanied by the Mohammedan whom I met this morning while looking for 191 Cecil St. Andreas acted as interpreter and is going to try to get together a meeting of Mohammedans, to discuss the American mission and raise subscriptions. In the afternoon I called upon my old friend Ong Kim Cheow and after a short chat I took a stroll through the town and drifted into the Chinese quarter where a Chinese open-air theatre was in progress. Of

---

[175] The Raffles Hotel opened in 1887 and was named after Sir Thomas Stamford Raffles, 1781-1826, founder of modern Singapore. K. Mulliner and Lian The-Mulliner, *Historical Dictionary of Singapore*, 124.

[176] Should be Rounseville Wildman, 1864 -1901, a native of New York, served as American Consul to Singapore from 1890-1892. He was later appointed Consul in Barmen, Germany in 1893 and Consul-General for Hong Kong in 1898. He was lost at sea on a voyage to San Francisco in 1901. Wildman was author of several works, including *Tales of the Malayan Coast* (1899). Oscar F. Adams, *A Dictionary of American Authors*, 581; and Susan Holly, U.S. Department of State, Office of the Historian, e-mail message to editor, July 16, 2004.

course the whole thing was unintelligible to me and consisted chiefly of the banging of drums and gongs and the squawking of wind-instruments. The players were boys and they were dressed in fantastic costumes and occupied their time in hopping and twirling about the stage, brandishing knives and bludgeons and twisting their bodies into all manner of shapes. Their costumes were changed in full view of the audience and the stage hands and supers walked about among the players in the most matter-of-fact way imaginable. It became necessary in the course of the play, I suppose, for three of the actors to pass through a doorway. One of the stage hands held up a piece of cloth painted white and blue with an archway cut in it and the three actors calmly marched though it while the gongs and tom-toms set up a deafening din. The stage hand calmly folded up the cloth, threw it in a corner and went off to one side of the stage to regale himself with a cup of tea from the pot sitting on an open charcoal stove, while the actors went on with their antics. There is one thing about a Chinese theatrical performance that impresses me; it is the active and vigorous exercise it gives to the imagination of the audience. They must imagine almost everything except the actual presence of the players. Woods, castles, houses, plot all must be imagined. It must be a strain on the mind but the Chinese are used to it. I waited until there was a lull in the performance indicating a division of the play into acts and then I started on a stroll through the streets. It was evidently the Chinese quarter of the town for all the stores were kept by Chinamen, and the whole aspect of the locality was plainly and unmistakably Chinese. It was but a repetition of the scenes I have witnessed in Manila many times in the streets occupied by Chinamen. There were the various stores giving forth the same kinds of smells; the same children playing about the doors in the dirt and the same air of uncleanliness everywhere. Everything was the same except that here there is a scarcity of dogs—I may almost say that there is an utter absence of them for I haven't seen a dog since I arrived here. On the way home I stopped and listened to the band on the Singapore Cricket Club grounds and saw just a little of the game which, to my mind, is exceedingly tame when compared with our Base Ball. But the grounds of the club are grand. They have a most level lawn, perhaps two or three acres, on the sea shore with a fine two-story club house at one end and in the centre a statue of someone who looks, at a distance, like James O'Neil as the Count of Monte Cristo.[177] Around the lawn is a fine driveway which was liberally dotted with carriages containing ladies and children—all English, of course—and some wealthy Chinamen. The carriages are nearly all light, open affairs, something like a Victoria, drawn by one large horse. Its cheaper to keep one horse than two. The vehicles for hire are called "Garries" and are very easy, comfortable affairs. They have 4 wheels and doors on the sides with seats facing each other and are closed in by blinds to keep out the sun. These can be lowered leaving the body comparatively open. But the Tamil and Malay drivers are just as great frauds as the hackmen of any other country. I got one this afternoon to take me a short distance and under his tariff which was posted up in his "garry" he was entitled to 15¢. He insisted upon "Hap, dollar" and I had to call a policeman to get him to settle for 20¢. I had one four times as long this a.m. and paid 35¢ but the payment was made under the direction of the hotel clerk and the rascal didn't dare to demand more.

September 15, 1892

Arose at 5, bathed, prayed and started out to see the town. Took a jinriksha and, I suppose, am everlastingly disgraced in Singapore. English snobbishness prevails here to even a greater extent than in Manila. In Manila Englishmen will not ride in street cars nor carromatas because it might give the public an impression that they were not wealthy and aristocratic. But they will sometimes ride in these vehicles at night and everybody knows it. Some of the clerks—those who are not so overwhelmed with debts that they can keep a cheap trap,—will not ride in the cars or carromatas at all. In Singa-

---

[177] James O'Neil, 1846-1920, Irish born actor who became famous for his role as the Count of Monte Cristo, playing the part more than 6,000 times over three decades. *American National Biography*, 16:731-32.

pore even the English who cannot afford to keep traps, will not ride in the street cars at any time and in the jinrikshas only at night. It would be considered very detrimental to one's social standing to ride in one by daylight. And yet they are the most comfortable little vehicles imaginable,—clean (as a rule) and as easy as a rocking chair. But they are cheap, and the average English tramp hates bitterly to have anyone know of his poverty. I passed some beautiful residences on the hills and rolled along some of the most attractive drives I have ever seen. A landau has just passed the hotel in which is seated a young Englishman driving while a Tamil servant perched up behind holds a white sunshade over him. Now it wouldn't surprise me to learn that at home in England the fellow went to his work in a cheap omnibus and, perhaps, carried a dinner pail. It is astonishing how quickly these people become Lords and Dukes—in their minds—when they get out in the East where servants are cheap and it is not considered proper to let the natives know that they had to work "bat 'ome in hold Hengland." They make me very weary sometimes. Tamil laborers as black as your hat and entirely naked save as to a breach cloth tightly wound about the hips, and a turban, work on the streets, drive ox-carts and garries and seem to be as industrious as the Chinese. Spent most of the morning at the U.S. Consulate and took tiffin with the Consul and his wife at the Hotel del Europe. In the afternoon went to the "Gymkhana" at the Race Course, given by the Sultan of Johore in honor of his guest the Sultan of Pahang.[178] I was presented by the Consul to both Sultans. The Sultan of Johore, who speaks English quite fluently and dresses in English style is an independent sovereign who rules over the province of Johore, just across the strait from Singapore. He is wealthy and affiliates very readily with the English who make much of him. At the track I was also presented to Sir Charles Warren an English military man who

---

[178]Sultan Abu Bakar ibni Al-Marhum Temrnggong Ibrahim (1833-1895) ruled Johor under various titles, including Sultan, from 1862-1895. Sultan Ahmad ibni Al-Marhum Bendahara Ali (1836-1914) ruled Pahang under various titles, including Sultan, from 1863-1914. J.M. Gullick, *Rulers and Residents*, 355-56.

71

had distinguished himself in Africa and other parts of the East.[179] He proved to be a very intelligent and entertaining old gentleman. The "Gymkhana" proved to be a series of scrub races for ponies, roadsters and garry ponies, for which prizes were offered by the Sultan of Johore who paid for the wine and refreshments dispensed. It must have cost him a very decent sum. There were present about 200 English men and ladies and about twice as many Malays, Tamils, Klings and other races. His excellency, Major-General, Sir Charles Warren, R.E.—G.C.M.G., K.C.B. is in command of whole infantry force.

September 16, 1892

I had a strange and impressive dream last night. I dreamed that Mamie and Nala went away from the house to play and remained so long that we were alarmed.[180] Presently Mamie returned alone and said that they had been playing on the bank of a pond, that Nala had fallen in and had disappeared. The poor little thing was very much frightened and I was almost wild with grief. My expression of grief awakened me and I was glad to find that it was only a dream. God grant that nothing serious has happened to my precious darlings. This morning I arose as usual and after bath and prayers strolled up the Beach Road in front of the hotel; visited a Buddhist temple of elaborate design but there were no prayers going on. There were two or three small joss sticks burning outside and as many in the space before the altar. All the elaborate belongings of a Chinese Buddhist temple were there and on each side of the altar stood a cheap American

---

[179] Sir Charles Warren, 1840-1927, a native of Bangor, Wales, served in Palestine, Australia, and South Africa. Acted as chief-commissioner of the London metropolitan police in 1886 and resigned in 1888 after several negative incidents, including the unsuccessful investigation of the Jack the Ripper murders. He served in command of Singapore from 1889-1894. Cyril Bailey, "Warren, Sir Charles (1840-1927)." in *Oxford Dictionary of National Biography, 1922-1930*, ed. J.R.H. Weaver (Oxford: Oxford University Press, 1937).
[180] Mamie was presumably a nickname his eldest daughter Mary, and Nala was his youngest daughter.

clock of the old gothic pattern. Took a jinriksha and rode up the beach road to the market; returned by another street. Saw nothing worth recording. There were the usual Chinese shops and the procession of races with their native costumes which are a never-failing source of interest to me. The gharry driver too is a most interesting creature upon the same principle that an American hack-man is interesting; no matter how much you pay him he insists upon having more and is never ashamed to ask for it. Yesterday one of them drove me half-a-mile for which, under the tariff card posted in his vehicle he was entitled to 15¢. I gave him 20¢ and he coolly demanded "hap-dolla."

This afternoon received an invitation from Sir Charles Warren to dine with him Sunday evening and accepted it.

This afternoon Ong Kim Cheow called about 3 and we went out in a gharry to visit some of the temples. First called at the Chitty, Hindu temple. The Chitties are Hindu money lenders. This is probably the most elaborate of the Hindu temples in Singapore and its walls are covered inside with paintings of the gods Brahma, Vishnu and Shiva and their satellites in various groups—some in chariots, on thrones, on horses and in numerous fantastic designs. Inside the main temple or court was a smaller temple very fancifully decorated with lamps and colored globes. Nothing but cocoanut oil is burned in the lamps and on the pavement outside the inner temple were piles of cocoanut meat drying; from this the oil is tried by heat and compressure. I was permitted to enter the inner temple upon taking off my shoes. At its rear end was what appeared to be a stone vault with heavy brass doors to which were hung about two dozen brass bells. The casement was also of brass with little shelves or wells up and down the sides in which was cocoanut oil and lamp wicks. The doors of the vault were closed and I was told that there was a beautiful image of Brahma inside but that it could not be seen until the doors were opened for the regular service at 4:30. It was then only 3:30 and we determined to visit a Mohammedan mosque and other places and return in time for the Hindu service. We found the largest Mohammedan mosque but were only allowed to enter the first room, and then with my shoes off. The old man who seemed to be the chief of the attendants was apparently very suspicious of me and would not allow me to enter the prayer room. He said the chief man was not in and he had no authority to admit strangers. We then went to another and a smaller mosque where we were not only admitted but were very cordially received by the Moulvi and several of the attendants and were prayed for in Arabic. Incense was burned and lights were lighted behind a veil which seemed to be a sort of altar. Other lamps were in front of the veil. All this, of course, was a ceremonial not warranted by the teachings of Mohammed. At the conclusion of the prayer the Moulvi touched us with a bunch of peacock feathers as a blessing or benediction and gave us a charm in the shape of a few flowers and a bird done up in red paper. Where these Moslems find a warrant for such things I cannot imagine. It is evidently a case of gone astray. We then went to a handsome Chinese Buddhist temple near by and were allowed to roam about it at will, shoes and all. It was the most elaborate building of the kind I have yet seen and contained beside the goddess of the Sea, the clairvoyant and clairaudient gods. A yellow robed priest was performing a service before the altar while a male and female attendant were lighting joss sticks and making ready for the sunset service. The main temple and surrounding buildings were very elaborate specimens of Chinese architecture. We would have liked to stay longer but wanted to get back to the Hindu temple in time for the 4:30 service and it was then 4:15. When we arrived at the Hindu temple a company of Bengalese Hindus were squatted in one of the side sheds, listening to a yellow robed priest while he chanted in drawling monotones a prayer which was occasionally varied by the ringing of a bell, and the blowing of a blast on a horn. The priest was seated on a platform raised about two feet from the pavement and one of the listeners who sat in front of him made an occasional response and threw a few flowers on the platform. When we arrived, there were about a bushel of blossoms in front of the priest. Soon after another Hindu joined the group and began beating a tom-tom while the rest chanted a hymn. We turned to look for the man who had admitted us to the temple on our first visit and failing to find hind him we asked two others when the service would commence in the temple and were surprised to learn that it would not take place until 6:30. Another man said 5:30 and another 6. Then we encountered the Hindu who had given us the hour as 4:30 and he amended his statement so as to read 5:45. Of course we were discouraged and disgusted as it was evident that the people did not want us to witness the service. We then started for the Shrine of Iskander Khan which is on the southern slope of Fort Canning Hill near the old cemetery. We found it a low stucco-covered tomb enclosed by a high wooden railing with an opening at one end through which visitors make their offerings of fruit, flowers and sweetmeats. A vase of incense is kept burning at the foot of the tomb day and night. It is supposed to be the tomb of the Sultan Iskander, or Iskander Shah, which is the Eastern name for Alexander the Great. and it is considered a very holy place by the Mohammedans of Singapore.[181] Who believe that it is several thousands of years old. Its history seems to be shrouded in mystery and uncertainty and there are many conflicting legends concerning it. It is said to have been discovered by accident after the British settlement of the island in 1819 when the jungle on Fort Canning was being cut away. It is difficult, therefore to say how the Mohammedans acquired the idea that it was the tomb of a Mohammedan prophet as some of the Moslems here evidently believe it to be. On Friday and Sunday evenings the shrine is visited by many Mohammedans who believe that their diseases will be cured thereby. We found attached to the railing around the tomb numerous bits of tile and stone suspended by strings and supposed to contain the diseases left there by the suffering pilgrims. I found four half naked Arabs there who readily admitted me to the space around the outside of the railing and gave me some incense to put on the fire and some fruit, flowers and sweetmeats to offer at the foot of the tomb. When I left I gave one of the men 20¢ which he took me-

---

[181] Iskander Shah was a late fourteenth-century Malayan ruler who converted to Islam and eventually seized power on the island of Temasek, which about that time was renamed Singapura (Lion City). Ernest C.T. Chew and Edwin Lee, eds., *A History of Singapore*, 5.

chanically and I heard him and his companions talking loudly as I passed down the hill; whether they were quarreling over possession of the money or anathematizing me for not giving more I cannot say. The tomb is most delightfully situated under an immense tree which completely shades it and on the side of a hill that commands a fine view of the surrounding country. We visited another Hindu temple which was being repaired for a coming festival and saw some of the faithful repairing and repainting the gods. I was not allowed to enter without taking off my shoes although there was a fine looking heifer calf tied to one of the pillars inside the temple and the floor was littered with dirt and the refuse of the carpenters and plasterers at work there. In one of the side rooms were 7 or 8 children, almost naked, seated on the floor in a line repeating parrot-like their lessons for the afternoon while the schoolmaster strutted around with a cane in his hand large enough to have knocked a horse down with. One of the little ones sidled up to me and opening my hand put two or three flowers in it and looked up at me with a most engaging smile. It caught me for 2 cents. Immediately another child slipped out of the line and hurried out to get some flowers. He caught me later on. Two big Hindoos asleep on the pavement. Ong Kim Cheow then took me to the house of a wealthy friend—a regular Chinese house. We passed through an arch-way ornamented with the usual dragons and other figures, into a garden filled with rare plants and with two large aviaries containing a variety of beautiful birds. Some of the plants were trained over wire forms to represent birds, beasts and fishes as well as human forms. We passed into the house proper in the centre of which was a court open to the sky, and containing some of the most beautiful ferns I have ever seen. There were also pieces made of rock with earthen ware figures of Chinamen, representing a Chinese house or garden on the side of a mountain. Miniature trees grew from the clefts of the rocks—actual living trees—or rather, small plants resembling trees and most of the places were covered with fine moss. I never saw anything in miniature so life like. It was an animated domestic scene on a mountain side redressed to the last possible dimensions. The house was elaborately decorated in the Chinese style and the walls were covered with pictures and works of Chinese art. In one of the side rooms was an elaborate and richly decorated altar upon which were deposited the ashes of some of his ancestors. Of course we did not see the owner's wife or wives; they were in a part of the house which a stranger is never allowed to enter. A Chinese wife cannot go astray even if she wants to. In the evening Ong Kim Cheow escorted me to a part of the Chinese quarter where the festival of Seven Moons or Sambayan was being observed. This is observed in September and has special reference to the propitiation of the boss devil and his satellites. At some time during the month each street or district must have it's houses decorated with colored paper lanterns, fantastic figures and lights, fruits, sweets, food, tea &c are offered to the devil and prayers and chants and the beating of tom-toms, and wooden pieces, the squawking of wind instruments and the wails stringed fill the air. Open-air theatre give performances, and a general air of festivity prevails. This evening the celebrating district was South Bridge Road and Canton and an adjoining street. The scene was the most animated of its kind I have ever seen. Prostitutes on Canton and Hong Kong streets.

September 17, 1892

Was out before six this morning and visited Arab street where there are many Arab tradesmen. Strolled through their shops and bought a cap for 20¢. Returned to hotel and found that the smell of a dead animal which I noticed slightly yesterday had increased to such an extent that I could not endure it. Reported the matter to clerk—stupidity of Chinaman who insisted upon pouring water under house and washing steps. Clerk called a Malay boy—no better results—changed to much better room. Sparred awhile with Hindu boy who wanted to sell me a Sarong, Sash, and other Indian goods. Didn't buy anything—too poor and have wife and three nice children to look after. "Do I go away empty?" asked the Hindu sadly. Spent an hour in the Raffles Library and Museum on Orchard Road.[182] It has an air of newness about it that is discouraging to the seeker after novelties and it is not well-kept. There is quite a large collection of fishes and reptiles but the specimens are, many of them, badly mounted and there is an exasperating absence of descriptive labels. The specimens that are most apt to arouse curiosity of the visitor probably will have no labels. There is quite a fine collection of Malay and Indian boats but, like the fishes, they are not arranged and kept in an orderly manner, and are not shown to advantage. There is quite a full collection of the birds of the Malay archipelago and a very small zoological collection. The library seems deficient in the higher class of literature and liberally supplied with popular novels and the current humorous periodicals and English dailies. The Logan Library which I found specially mentioned in the new "Hand Book of Singapore" and admission to which could only be obtained by special permission from the Librarian, was a great disappointment to me. It was said to be composed of Oriental literature and I naturally acquired the impression that it contained some of the Eastern religions or mystical works—but not one was to be found. There were perhaps a thousand books all told and they treated generally of the Oriental languages, history, philology, zoology, ornithology, and several other "ologies" but there was nothing even on Oriental literature, sacred or profane. My visit was rather disappointing all around. Returned to the hotel, tiffined and spent the afternoon in a long chair with the August "Theosophist" as a companion.[183] In the evening Ong Kim Cheow escorted me to a Chinese club-dinner and general recreation, as he called it where I met two other Chinamen who spoke

---

[182] The Raffles Library and Museum Opened in 1887. C.M. Turnbull, *A History of Singapore, 1819-1988*, 110.

[183] The *Theosophist* was founded in 1879 as the first of many publications of the Theosophical Society. Marion Meade, *Madame Blavatsky*, 208. Webb kept current with Theosophist literature, believing he would someday convince them to become Muslims. He wrote in July 1892: "...to be a Theosophist, one must be a follower of Islam." Webb, *Islam and Theosophy*, 421.

English fluently.[184] These gatherings which form the popular amusement among the better classes of Chinese—the wealthier classes of young men—are peculiar as showing one of the phases of the Chinese social system which has been very strongly condemned by the Anglo-Saxons. Each man pays $2 which entitles him to his dinner and the companionship of one Chinese girl during the evening, who sings for the benefit of the whole company and plays on a sort of zither, if she has learned to, (and most of these singing girls have) The men assemble in a large room with seats around the sides, couches or benches for opium smoking and several tables, large and small. Drinks of all kinds are freely dispensed including brandy which the English have taught the Chinaman to drink, and during the early part of the evening the men lounge about and "monkey" with the girls while the latter sing to them to the accompaniment of the shrieking Chinese fiddle and banjo. The girls, the majority of whom are quite pretty—even beautiful from the Chinese standpoint—are procured in China, either by purchase from their parents or by abduction and are destined to become either prostitutes or concubines,—they can never be anything else. They are owned, literally, by women who bring them here, train them to sing and play the zither, rent them out to sing at gatherings of this kind and ultimately sell them to a Chinaman who wants a mistress, or put them in houses of prostitution. One reason why they are sent to these social gatherings is that they may exhibit their charms and accomplishments and attract a purchaser. While the men dine these women wait upon them, fan them and look after their comfort generally. Beyond fondling and, perhaps kissing and joking with them nothing out of the way occurs at these gatherings. The girls are guaranteed to be virgins and there is very little more freedom between the sexes at this time than there is usually at a gathering of Christian young people. But just enough liberty is allowed to inflame the passions of the men and one or more sales of girls follows, sometimes. If a man becomes

---

[184] The club was likely the Straits Chinese Recreation Club, of which Ong Kim Cheow was one of five founding members. Ong Siang Song, *One Hundred Years' History of the Chinese in Singapore*, 216.

enamored of one of the girls he ascertains who owns her and makes an offer for her. If she is specially attractive the price may be held up to one or $2000 or even more. If she is older than she ought to be to have a high market value she may bring only $250 or $300. As soon as the purchase is made the man takes her to his house and there she remains as his mistress. She may become his legal wife if she bears him children. At all events she disappears from the concert room—or club-room—and escapes a life of common prosti- tution. If she cannot catch a purchaser, she must become a prostitute. Her female owner brings her to these club dinners and escorts her home again seeing to it that she is not led astray for in her virginity lies her chief value to her. Kim Cheow, another Chinaman and myself left the gathering about 9 and went to see a "Sambayan" in an adjacent street which was reported as unusually brilliant. It so proved. There were three long tables at least ten feet wide set at one side of the street and forming a long table fully 150 feet in length. This was filled with food of all kinds arranged in the most attractive manner possible and decorated with colored paper flowers and Chinese ornaments of various kinds. There were also numerous candles, and bunches of gas jets at intervals along the sides of the tables made the street as light as day. All this was done to propitiate the devil and his angels. At the head of the table was a carpetted bench where the old 'un and his friends could sit if they wanted to accept this feast, and a wash-stand and towels, where they could wash their hands before eating. The first attraction on the table was a dozen bottles of champagne as many of beer and others of rice wine, sherry &c. with glasses. Then there were about fifty made dishes of food the composition of which I know nothing. Chickens dressed like men. Then came whole roast pigs and fowls and fruits and sweet-meats, and cakes in an infinite variety. I am sure there was food enough on that ta- ble to satisfy two thousand people. All this was offered to the devil by the Chinese shop-keepers of the street in the hope that they would thus gain his good will so that he would not interfere with their affairs during the ensuing year. On the street also were Chinese theatres and bands in full blast, all at the expense of the residents. After the offering, the devil having failed to eat up the food it is distributed around among the friends and relatives of the donors and what is left after that is given to the poor. This month the devil is fed most sumptuously by the Chinese, but the rest of the year I rather think he has to hustle around and get his own feed. The junks, sampans and barges were also celebrating last night but their performances I think consisted principally in the illumination and decoration of their boats, the offering of simple food to Satan and the banging of gongs and tom-toms to keep the little imps from filching the big 'un's food.

September 18, 1892

I was up so late last night that I did not arise at 4:30 as usual but prolonged my slumber until 7. This is a typical Sunday morning—quiet, the birds nestling among the flowers and trees in front of hotel &c. Went to the Botanical Garden and, as a result, made a discovery of a new method of robbery. I took a gharry at 9:30 and returned at 12. Distance 4 miles—time 2 ½ hours. When I returned to the hotel I asked one of the clerks how much I ought to pay the gharry-man. He discussed the matter with a man at the meet desk and he decided that I ought to pay $1.50 that the driver would probably ask more but that $1.50 was enough. I handed the driver $1.00 and he turned away looking at the clerk in rather a disappointed way. He hadn't heard the conversation. Then an idea struck me. I remembered that last week one day I asked one of the clerks how much I ought to pay the gharry-driver and he said 35¢ which was nearly double the amount called for by the printed legal tariff card. I saw by the printed tariff card also that the tariff price for a gharry for 8 hours, and a distance not over 10 miles was $1.25 and yet the clerk said I ought to pay $1.50 for 4 miles and 2 ½ hours. As these gharries are about the hotel all the time the possibility suggested itself to me that the clerks might have an interest in assisting the gharry-drivers to rob guests. This is only a surmise but it seems to have a sort of foundation in fact. But to the Botanical Garden. I was driven over some of the most nearly perfect drives, in order to get there that I have ever seen. They were broad, smooth and hard and shaded by immense trees.

On earth side nestling in the jungle on the hillsides were beautiful country residences with fine driveways leading up to them from the main road. What a relief and pleasure it was, after a five year sojourn in dirty, ill-kept Manila, to see such magnificent drives and such fine residences! It confirmed me in the belief that Singapore is one of the most beautiful and attractive of cities. I strolled through the Botanical Garden and was impressed by its future possibilities. It is evidently young yet and cannot justly be compared with the older ones of Europe and America. But it is beautifully laid out and contains a large collection of rare trees and plants, a very pretty lake upon which ducks and 3 black swans are kept, and a small collection of animals and birds. It is hoped that a fine zoological collection can yet be had, but new cages are badly needed for those already on hand: 2 bears, 2 deer, some squirrels, monkeys and birds. All the cages except those for the bears and monkeys are very cheap and badly kept. Names of plants and trees rarely given except in Latin. On the way home met a Malay funeral, the corpse on a stretcher and covered with a closely sealed arched covering of bamboo covered with colored shawls. Asked the gharry-driver what it was. He was able to speak some English. Said it was a Malay funeral. As it turned up the first road after passing us the idea seized me that I would like to see a Malay funeral service and I asked the driver if it was very far away where they were going to bury the man. He said: "Yes, very far." "Well let us go any how" I said.

"Oh no, no can, Malay stop carriage. No let Master see funeral. No like it. Chinaman yes, Hindu yes; but Malay no. Mussulman no." Thus I was deprived of the pleasure of seeing another novelty. But I am firm in the belief that the gharry-man lied and that he had some private reason for not going near the funeral. Spent the afternoon in my long chair and in the evening went to dine with Sir Charles Warren. Consul Wildman sent his carriage for me and I had a most delightful ride in the cool night air. Sir Chas lives in a most attractive location on a hill surrounded by trees and calls his place "Balmoral". The company consisted of Sir Charles, Mr. and Mrs. Wildman and a small, bald, Psychical Researcher named Ridley, the superintendent of the Botanical

Garden.[185] As he was an upholder of the Hodgson Investigation fiasco and an opponent of the T.S. we didn't get on well.[186] He claimed an extensive knowledge of psychic matters and a vast experience in thought transference. He tried two experiments in thought transference, both of which failed and then gave it up. I claimed to know all about Theosophy and the people at the London Headquarters, and turned up his nose at them. He was one of those self-sufficient, and very gullible people who always proceed on "scientific lines" and (in their minds) are never "taken in". Taken all around the evening was not a very pleasant one. Arrived at hotel about 11:45.

September 19, 1892

Didn't get up until 6:30, too late for a walk. The day was principally put in in sight-seeing, or rather, in trying to see sights—nothing of unusual interest was encountered. Kim Cheow promised to come up in the evening and escort me to a Chinese spiritual seance but failed to connect. Therefore I started out on my own account in a jinriksha. How in the world these jinrikisha men manage to hold out I cannot imagine. They will trot along mile after mile at a good run hauling their vehicles and never stopping to rest or take

---

[185] Henry Nicholas Ridley, 1855-1956, born at West Harling Hall, Norfolk, England, and came to Singapore in 1888. Acted as director of the Singapore Botanical Gardens from 1888-1911. He is often regarded as the "father" of the rubber industry for discovering a means of tapping trees without harming them. Mulliner and The-Mulliner, *Historical Dictionary of Singapore*, 128-29; and Edward J. Salisbury, "Ridley, Henry Nicholas (1855-1956)." in *Oxford Dictionary of National Biography, 1951-1960*, eds. E.T. Williams and Helen M. Palmer (Oxford: Oxford University Press, 1971).

[186] In 1884 Dr. Richard Hodgson, 1855-1905, was sent to India by the Society for Psychical Research to investigate the spiritualist claims of the Theosophical Society's founder Helena Petrovna Blavatsky. The resulting report concluded that the claims of the Theosophical Society and Blavatsky were false and fraudulent. Arthur S. Berger and Joyce Berger, *The Encyclopedia of Parapsychology and Psychical Research*, 181-82.

breath. They go to Johore, 14 miles, for 70¢, up hill and down, and are as fresh as a horse would be going that distance. I found my way to Malay Street and witnessed a scene unequalled by anything I have ever seen in an American city. Hundreds of women of all nations soliciting from the verandas of the houses. Japanese predominating. In one of the side streets saw the Chinese ceremony over the dead. The corpse was that of a woman, probably a prostitute, who had been dead but a short time. It was laid out on a low bench covered with its clothing and with a paper mask over the face which was subsequently removed. It was literally covered with prayer papers and at each hand was a bag, supposed to contain coin to defray the expenses of the soul in the next world. At the foot of the bench was a light burning and several small dishes containing food,—offerings to the devil. The general idea seems to be to distract the attention of the devil so that its will not try to catch the soul of the departed nor interfere with its progress in any way. To this end, it was that three priests chanted prayers, and banged gongs, bells and wooden tom-toms, while a paid musician squawked vigorously on one of those shrill-voiced Chinese flageolets. Two women were busily engaged in burning prayer papers while three hired mourners moaned and groaned and wailed most agonizingly. They earned their wages. Then the priests performed a sort of mass before an altar at the back of the room and the musician, the same person I saw the night before leading the music at the Chinese club dinner, changed his flageolet for one shriller in tone and proceeded to torture the air with its strains. This sort of thing, with slight variations was kept up until midnight when the soul of the deceased was supposed to be fairly on its journey, and the priests, musician and friends retired leaving the hired mourners to watch the corpse and keep the candles lighted. North Bridge Road—Chinese Christian chapel—Union Hotel and street scenes.

September 20, 1892

Arose early and went out for a walk. Group of Chinese boys playing marbles in Middle Road—like American boys ex-

cept that in shooting they held the marble between the thumb
and fore-finger of the left hand and shot it out with the middle finger of the right. Chinaman selling live frogs in
bunches. Chinamen squatting around dealer, eating small
clams, with corrugated shells, dipping them in a mixture of
condiments. Refreshment and food peddlers are numerous—
all Chinese. Go where you will you will see them and they
generally appear to be doing a thriving trade. Food is plenty
and for two or three coppers one can get a good hearty dinner. No one need starve here who can raise five cents a day.
The food is mainly meat and vegetable stews—Heaven
knows what the meat is. In Manila some Chinamen were detected in putting human bones into their soup—But there are
also pork, chickens, ducks, crabs, fish and various other
meats in a state that allows of their being examined and
judged upon as to wholesomeness. Ice cream and cool drinks
are sold as well as tea and coffee. Most of the drinks have
preparations of isinglass in them and the water is poured
over small cakes of ice set in a funnel shaped tin vessel suspended on an iron rod about two feet above the stand. The
street story-teller or reader is surrounded by 15 or 20 Chinamen—sometimes more—and reads a story to them occupy-
ing perhaps an hour. At its conclusion each listener pays half
a cent. Black, shiny, naked Tamil in the rain seated on his
bullock cart. Hindus with ashes of cow dung smeared on
their foreheads, breasts and arms, and the caste mark on the
forehead between the eyes. "Dubash" is a comprador or ship
chandler—"Samshoo" is rice liquor. While in a shop on
Beach Road to-day looking at some caps a Mohammedan
tailor came in. I had never seen him before. He looked at me
intently and then held out his hand giving mine a fraternal
grip. How he knew I was a Mohammedan I cannot tell unless
he had received a description of me from Andreas or someone else here who has heard of me. Spent the evening in my
room reading.

September 21, 1892

Arose at 4:30 and went out for a stroll and "riksha" ride, after having bathed and prayed. Saw nothing of interest except

a large gathering of Tamil and Kling street laborers waiting to be set to work. The morning air was damp and chilly and they were wrapped in their white and red clothes looking probably as their progenitors did hundreds of years ago. The fashions in the Tamil and Kling costumes never change. Closed up my packing to leave on the steamer "Patna" of the British India Steam Navigation Co. for Rangoon. Left hotel for Tanjong Pagar Wharf at 1:45—slow bullock cart wanted 10¢ extra for making bullocks hurry—boat loads of beautiful corral and shells—water very clear for a harbor—boys diving for 5 cent pieces—fragile canoes. As ship sailed out other canoes came about in deep water, dancing over the swells from the vessel. When I went on board found my cabin occupied by the baggage of a corpulent German tourist who gave it up with very bad grace. Two days before I had selected it because it was large and roomy. Steward said he would hold it for me but when "Deutchy's" baggage came he thought it was mine (he said) and had it put in. The aspect in passing away from Singapore and up the Straits of Malaca very attractive. Wooded hills dotted by villas and bungalows, and nipah roofed houses—native Malay houses on stilts over the water—excellent system of buoys and lighthouses—Johore, Selangore, Perak on the right,—Sumatra in the far distance beyond the small isles on the left. Water-closets all built for long-legged people. Passengers small in number but interesting. French Catholic priest who wanted to talk. "Parle vous Francais"—"No, Hablo Español y Ingles." Young Brit Lieut with tattooed legs came on deck in his bare feet and Sarong only to show his legs, apparently. Comparatively modest though for an Englishman. Tattooing looked like Jap designs—will try to find art. Company very quiet so went to bed at 8. Hindu pounded at door about 9 to shut port holes as it was going to rain. Like the "Patna" so far—big improvement on the "España". Capt. looks little like Capt. Peane.

September 22, 1892

Hindu pounded at door at 5 to tell me to shut port holes as it was going to rain. In revenge told him to arrange bath. Water very calm and sailing comfortable. According to log ran 110 miles during the night. Sailor men from Calcutta who speak Hindustanee—sounds a good deal like the old gibberish we used to call "hog latin." About sunrise we were opposite Malacca and passed Cape Richada about 9.[187] Very beautiful country with high mountain ranges inland. Country well supplied with wild elephants, tigers and numerous kinds of poisonous snakes. Sumatra in the dim distance on the left. Caught a distant view of Mt. Ophir which some claim is the Mt. Ophir from which the gold was brought to build Solomon's temple. The best butter since I left the States is served at table and the fare generally is good except the crackers which are mouldy. Passed out of sight of land shortly after noon. Commenced to read "David Grieve" in the afternoon and it made me homesick—I longed for my precious babies and went to bed feeling quite miserable.[188] I wonder if anyone grows happier as he grows older. There is certainly very little real happiness in this life. I look forward to meeting my babies and feel that I will be happy with them. But they will grow to manhood and womanhood and I will grow old and troublesome and perhaps will know that they will wish me dead and out of the way. Oh, well, God's will be done.

September 23, 1892

The Lieut hasn't paraded his tattooed legs with the enthusiasm he displayed yesterday morning. Perhaps their failure to create a sensation among the passengers has discouraged him. The priest seems blue and out of sorts and does not seem as anxious to talk as he did the night before last. He is German, I find, and he and Deutchy had quite a long confab yesterday. Perhaps Deutchy has acquainted him with his grievance concerning the cabin. Deutchy comes on deck occasionally now—took his tea there this morning—and seems to be rather more at peace with the world than he has been heretofore. The sick young man from Assam, who tried to draw me into conversation yesterday on the ground that he had lived at San Francisco and had been in Chicago seems to

---

[187] Should be "Cape Rachado."

[188] *The History of David Grieve*, by Mrs. Humphry Ward (1892).

have had a relapse. Oh yes here come the tattooed legs. I thought he was irrefusable. About noon we were abreast of the Island of Penang and were rounding its point to go to the city. We are nearly opposite the extreme point of Sumatra where the city of Acheen is situated.[189] After leaving Penang we pass into the Indian Ocean between the Nicobar and Andaman Islands on the left and the Malay Peninsular on the right into the Gulf of Martaban on which Rangoon is situated. We turned into Penang harbor about 1:30 and dropped anchor at 3 amid a heavy rain and dense fog. Of course we couldn't go ashore in a drenching rain and a wind that made the sampans dance at a lively rate. Soon after we had cast anchor a sampan came off to us with some Corinjie women from Coringa in the province of Godavery district of Rajahmundry on the sea coast north of Madras.[190] They had very few clothes on but jewelry enough to stock a small store. Each had a tiara of gold with a gold tassel or ornament hanging down over the forehead. Each ear had 3 pendants one in the lobe and two in the upper rims and in the nose were also 3 rings, one in the cartilage dividing the nostrils and one in each outside wall of the nostrils. On the neck were several necklaces and on each wrist a bracelet. All these were of gold on every finger and toe was a silver ring, and on the ankles were silver chains and bangles to the weight of two or 3 lbs I should judge. They chewed betel and their teeth were very much discolored but one of them was quite pretty and intelligent. They were accompanied by half a dozen half-naked men and came on board with 10 or 12 bunches of bananas, about as many green cocoanuts and other provisions to be consumed on the voyage to India. We also had two cases of cocoanuts come alongside for Rangoon. Later on a young man and his sick wife and 3 rather pretty children came on as 2nd class passengers. One was a dear little baby about 18 months old. I got hold of it of course and longed for my babies. Homesick again. We waited anxiously for the

---

[189] More commonly spelled Acheh.
[190] More commonly spelled Godavari.

rain to cease so that we could go ashore but it rained until bed-time. U.S. novelty Co on board. Appear to be broke.

September 24, 1892

The day broke comparatively clear and at 6:30, at the invitation of the captain, I accompanied him to the shore in his boat. Took a jinriksha and rode for half an hour through the residence streets. If I was charmed with Singapore I am more than charmed with Penang. In Singapore the roads are red, in Penang, white or gray. Beautiful level drives and handsome residences in groves of cocoanut palms. Cocoanut trees very thick and tall. Air from the mountains pure and cool. Discharged jinriksha in the retail quarter and strolled about on foot visiting the mosques and temples and peeping into the shops. Saw more Coringa women with enormous ear-rings and bracelets and anklets. Naked child with girdle of silver, silver ear-rings and nose rings. Saw silver anklets for sale at $12 and $15 per pair. Sign over Chinese Dr's office:

Ho Fat Cheong
Cantonese Doctor

Who cured the child of Mrs. Mahomed Arab,
when all hopes were given up by the European doctors. For truth of this see in the
Penang Gazette for 11th February 1887. This
board was presented by Mrs. Mahomed Arab,
Penang, 20th Feb., 1887.

Chinese woman with small feet got out of jinriksha and staggered into house. Weighed at least 175 lbs and feet not over 3 inches long. Another one—very old—toiling painfully along on feet of about same size. Horrible barbarity it must have been to her as a child. Sun getting hot, started for ship at 8:15 for breakfast. Sampan man wanted 20¢, regular fare 5. After breakfast at 9 went on shore and took gharry for the Botanical Garden and waterfall. The drive over the smooth white road through large groves of cocoanut palm was grand. Cocoanuts lying about the ground no one, apparently coming to pick them up. Botanical garden at foot of mountains, most beautifully laid out. Left gharry and had hot walk 1/3 way of mountain to foot of falls. Falls perhaps 100 feet high over almost perpendicular granite rocks, cracked and fissured. Water falls into basin on table and shoots off down ravine forming swiming pool in the valley below. On table is Hindu temple and plain wooden bridge across the ravine from which an advantageous view can be can be had of the fall and a glimpse of valley. The walk up the narrow winding path from the main road to foot of falls is steep but shaded and damp so that it is a relief from the heat of the garden below. The thick jungle and dank vegetation on the mountain side suggestive of snakes and tigers. Saw only 3 wild monkeys scamper off up the hill. Saw Victoria Regia— small specimen and not in bloom. Saw one black lizard about a foot long creep into a crack in the rock.[191] Grand Hindu temple built by Klings on road half way to garden. Klings abound in that section and seem to sleep most of the time. At wharf competition between sampan men and one agreed to take me to the "Patna" for 5¢—the first bidder wanted 20¢. My sampan man started for the nearest steamer and when out a little way I called his attention to the fact that he was not going to the "Patna", which lay further out. "20¢ to Patna". I told him to go ashore. Finally compromised on 10¢ Quite a number of deck passengers came on board, Chinese and Hindus. We raised anchor and started from Penang at 4. The view going out very fine.

September 25, 1892

Daylight found us nearly abreast of the Nicobar Islands on the left and the Mergui Archipelago on the right. To-night or to-morrow we shall be opposite the Andaman Islands on the left and the Tenasserim Provinces on the right. We are now in the waters of the Indian Ocean and sailing very smoothly. In fact there seems an air of quietness about everything. I have seen no evidences of marine life since I left Singapore and only one bird. The Sulu and China seas were full of it.

---

[191] Victoria Regia is a water lily.

90

Nothing eventful occurred during the day. In the evening we sat on deck—the Lieut, the Capt. and the Scotchman from Assam and told stories about tigers and snakes or, at least, they did. Capt. told about wood cutter bitten by cobra,—sent for snake charmer who could do nothing until he had seen the snake—found snake which bit charmer and both men died. Lieut told of 2 tiger cubs he found in Burmah, in rice store house of village. Villagers called him and told him tigers had been there but had gone away. Tigers never came back. Capt. told of pet alligator ship Capt. had—was showing it to friend and it bit off one of his fingers. Lieut told of army man in India out riding—tiger attacked his dog—ran and got riffle and shot tiger at once,—dog died. Capt. told of two English ladies out walking near Trincomalee. Tiger walked across road in front of them and disappeared in jungle. Lieut told of coolie in Burmah at work in field—tiger attacked him and seized left hand—battered tiger about head with big knife—tiger seized him by back of neck and still he cut and mangled tiger—next day coolie was found unconscious and tiger dead a hundred yards off. Coolie recovered but died afterward from effects. Capt. told of 2 Englishmen and 3 coolies out hunting tiger attacked and killed 3 coolies—1 Englishman ran away and the other stood still paralyzed with fear—tiger walked away without touching him. Scotchman told how leopard in Assam entered cottage and took child from mother's side and walked off with it.

September 26, 1892

Morning rainy and cloudy. There is very little to record. One of those sudden squalls peculiar to the Indian Ocean sprang up out of a calm atmosphere about 3 p.m. and whistled threateningly through the rigging, flapping the awnings and blowing the tops of the waves into the air in a creamy mist. It lasted for about half and hour and cooled the atmosphere very perceptibly. It rained nearly all day and we strolled about the deck, chatted and wished for meal-time. I put in most of the day stretched out on a woolen blanket on the sky-light, reading "David Grieve." Retired at 6:30.

September 27, 1892

Day opened warm and cloudy with the ship rolling rather uncomfortably. A rainy, squally, gloomy day with nothing to do but sit around and talk and read. In the afternoon the Capt. used Sir Wm. Thompson's sounding machine and explained its operation whereby soundings are taken by pressure at various depths. Lieut getting fresh.

III.

RANGOON

September 28, 1892

Daylight found us at the pilot ship, at the mouth of the Irawaddy River and at 6 we were steaming up toward Rangoon—46 miles—in charge of the pilot and escorted by a flock of sea-gulls, the first I have seen since leaving Manila.[192] The water of the river at this season of the year is very muddy and reminds me of the Mississippi. It is carried far out to sea discoloring the salt water for, sometimes, 50 miles. On the left we can see a long strip of land while the right is enveloped in mist and fog. Siam is invisible. Passed Elephant Point on left where British first landed when they invaded Burmah—marched from there 30 miles to Rangoon—tall stone obelisk to mark something which not even Lieut knows. Fish-hawks and gulls catching shrimp and fish turned up by propeller. Arab dhows,—ancient looking crafts anchored in stream—turbaned capt and crew. Signal station with cluster of native huts around it. Herd of buffaloes just like those of the Philippines. Burmese fishing village with peculiar shaped pagoda. Passed the B.I.S.S. "Madura" going out with Calcutta mail. River perhaps a mile wide in some places and twists and turns. Rangoon on the right and then

---

[192] More commonly spelled Irrawady River.

on the extreme left. Monkey point of the right. Burmese men tattooed from knees to waist and women made attractive to prevent sodomy. It was about 10 o'clock when we arrived in Rangoon River, or rather, that part of the Irawaddy which forms the harbor of Rangoon. As we neared the wharf I noticed a large number of Mohammedans awaiting our arrival but did not suppose they were waiting for me. But as soon as the gangplank was put off quite a number of them came aboard and I was soon surrounded by a crowd of Moslem brethren who seized my hand as fast as I could take theirs and gave me a most cordial welcome. I was not prepared for such a reception and my astonishment increased when I found that there were fully 500 Mussulmen on the wharf waiting to receive me. As soon as I had made ready my luggage I was escorted to the wharf where I passed through a long line of Mussulmen all anxious to take me by the hand. After them were two lines of school children who sang a song of welcome in Hindustanee. The reception showed a true spirit of brotherhood which was very pleasing to me. I was escorted to a handsome brougham and as we were driven through the streets to the principal mosque the carriage was followed by a crowd and eager faces peeped in at the window to catch a glimpse of me while many reached in and shook hands with me. We drove directly to the principal mosque where I addressed the vast gathering and one of the Moulvis responded welcoming me to Rangoon and congratulating me upon having become a follower of Islam. We then reentered our carriage and accompanied by the chief Imam and Abdul Careem Abdul Sakoor we were driven rapidly over very fine roads and through a most attractive landscape to the garden house of Moolah Ismail in Kemindine about three miles from the business centre of Rangoon.[193] The house which was large and cool and nestled among great

---

[193] Abdul Karim Abdul Shakoor Jamal, 1862-?, a native of India, he moved to Burma as a child. His family was engaged in trade and in 1884 he took control of the business, overseeing expansion into cotton, oil and saw mills. He was a well-respected philanthropist and acted as an agent for Webb's *Moslem World* publication. *The Cyclopedia of India*, 3:443; and Webb, *Three Lectures*, back matter.

94

Jack fruit and other large shade trees had been newly furnished and fitted up especially for my use and I was expected to remain there for about 15 days. I was the sole occupant of the house except the servants who had been secured for me. Soon after our arrival about 30 Moulvis and prominent citizens arrived and a sort of levee was held in the parlor, after which we had dinner on the ground floor after Mohammedan style. After that we all prayed together, led by the Imam of the mosque and then I stretched out in a long chair and had a needed rest. During the afternoon and evening I had a number of callers but was allowed to retire at 9. I was very much fatigued after the excitement of the day, and was glad to be alone. I did not expect such an outpouring of fraternal feeling and generosity from the Moslems of Rangoon and was deeply impressed by it. The crows very numerous and persistent.

September 29, 1892

Was up at 4:15 had a good bath and prayed. Spent the morning in writing to my loved ones. About noon had a nap and was about to pray when Bro. Abdul Careem Sookar and two other brethren arrived and we had prayers together. We then had tiffin and chatted nearly all the afternoon it being too rainy and muddy to go out. About 4 the sky cleared a little and after prayers Abdul Careem and I took a gharry and drove down to the Post Office where I mailed my letters. The P.O. attracted my attention owing to its convenient arrangement and the handsome appearance of the building. All the government buildings here have a solid substantial look and seem fully as good as those at Singapore. We then called at the office of Bro. Dawood where we were cordially received and I was given a chair while Bro. Abdul seated himself on the cushioned floor as is the custom here.[194] In all the offices of the Mohammedans there is at one end of the room a mattress or pad laid on the floor close to the wall, with cushions

---

[194] Bro. Dawood is possibly Dawood Ebrahim who latter corresponded with Webb. "News Notes," *Moslem World and Voice of Islam*, November 1895, p. 3.

for the sitters to lean back against. In front of them are boxes or desks upon which to write or count money. They seem very comfortable seated in this way and are so accustomed to it that they prefer it, decidedly, to chairs. We next went to Bro. Abdul's office and sat there for a few minutes after which we drove home. In passing through the streets I caught glimpses of Burmese life and customs that were interesting. While in a general way, the street scenes in Rangoon are very like those in Singapore and Penang, there are different features which are purely Burmese and which I hope to see more of before I leave. There are no jinrikishas here, the gharry being the popular conveyance. The Burmese seem to be a happy-go-lucky people and many of them are singing loudly as they pass up the road in front of my bungalow. To-day two funerals passed the house and they were a novelty. Line of ten ox-carts—tom-toms, pipes and cymbals, in carts,—catafalque of gilt and colors carried by 20 men, laughing and chatting as if going to a pic-nic. Women and men gaily dressed and carrying the inevitable Jap umbrellas following the carts on foot. Carts supposed to contain effects of deceased to be given to the poor. The second funeral had a black hearse and there were some signs of mourning. Seen to a disadvantage from the window, hope to see another at close range. Had the usual number of visitors during the evening all anxious to have a look at the American Moslem. About 10 some of my visitors said I might go to bed if I wanted to. I did so and left them in charge of the house. They chatted for about half an hour longer and then left.

September 30, 1892

Rain, rain and still more rain. The air is full of moisture and everything about the house is damp and sticky. This must be a fine house in dry weather, but, shut in as it is by the trees the sun has very little chance to get to it and it is very damp in the rainy season. And Oh! The crows! They are the most impudent, audacious birds I ever saw. They seem to be afraid of nobody and fairly haunt the houses flying in at the windows and helping themselves to anything eatable they can find. Yesterday morning one of them flew into my parlor and stole a piece of bread off the centre table that had been put there for my breakfast. He came in again this morning—or one of his chums—but he found no bread and went out disgusted. They sit in the trees close to the house and even perch on the verandah railings staring in at the open doors and windows, "cawing" in their melancholy way. Their extreme tameness and sociability is due, I think to the Buddhist regard for life which prevails. This is a Buddhist country and they take no life. The birds seem to have found this out and presume upon it. Yesterday I saw some white crows that frequent marshy places and are unapproachable in the Philippines; here they were strolling along the roadside as indifferent to the passing pedestrians and vehicles as if they were not there. Even the mosquitos sit still and allow themselves to be mashed.

I miss my loved ones more and more every day, it seems. I can hardly become reconciled to this separation. God grant that they may be kept alive and well until I return to them. They are my life and soul or part of me.

About 10:30 in company of Mahommed Dawood I took a gharry and drove to the city, this being the Moslem Sunday and special services being held at the mosque. We went first to Abdul Careem Sakoor's office and from there to the principal mosque where there were at least 1,000 Mohammedans already assembled. The service was very impressive—the grandest, I think, I have ever seen anywhere. To see 1,000 turbaned people bowing and praying in unison under the leadership of the venerable Imam, is a sight that no one can witness without a thrill of reverence and admiration. The Koran was read first by the Imam who stood on the platform with staff in hand and flowing white turban, presenting the appearance very similar to pictures of Arabian teachers that I have seen. After the prayers a simple funeral service was held over the remains of a young girl, and then I was requested to address the assemblage. But first I was formally given my Mohammedan name "Alexander Russell Abdulla Webb" to be written "A.R. Abdulla Webb."[195] I then addressed the assembly at some length upon the proposed

---

[195] Webb never used the name Abdulla.

American Propaganda, my remarks being translated by one of the brethren. This was responded to very eloquently in Hindustanee by one of the Moulvi's and there was much enthusiasm manifested. When I started to leave the mosque the people crowded about in great numbers to shake my hand and express their gratification at my having adopted Islam. Bro. Abdul Careem, Abu Beker and myself then drove down to a room that Abdul had engaged for me in the heart of the city and number of the brethren came to pay their respects to me. We had tea and fruit, and smoked the "hubble-bubble" and later on had ice-cream and jelly. There were two dishes and I was expected to eat both while all present looked on. This arrangement didn't suit me and I induced one of the others to take one dish. He ate part of it and gave the rest to a young man who sat opposite to me. At 4 we went to the mosque again to pray. This time the prayers were short. We then drove out home and had a chat and supper there. Nothing worthy of record occurred; but I was struck with the brotherly way in which the Mohammedans treated each other and the confidence they seemed to have in each other. The manner in which they have provided for and entertained me is far better than I could have expected or asked for. A well furnished country house, with servants and every convenience, placed at my disposal and every desire gratified as soon as my wish is expressed. The memory of this generous treatment will always remain with me.

October 1, 1892

Arose at 4 this A.M. and after bath and prayers went out for a walk up the road. Found a Burmese village the inhabitants of which seemed to devote themselves to the construction of Buddhist idols. In their little grass and leaf-thatched houses were idols of all sorts and sizes, from common clay up to richly-gilded and solid marble Buddhas seated in the conventional attitude of contemplation. These rich and expensive figures were in striking contrast to the squalor and dirt which enveloped the houses. I am inclined to believe that Burmese are fully as dirty and dreamy as the natives of the Philippines and that they are pure Malays. About 10 Abu Beker came and informed me that the steamer "Africa" on which Bro. Abdulla Arab was to arrive had been signaled and would soon be at the wharf. After breakfast we started off and reached the office down town just as Abdulla alighted from his gharry. The dear old man looked as loveable and good as ever and it was a pleasure to meet him again. He is far and away the best looking Mohammedan I have yet seen. We chatted at the office with the other brethren and then he went off to write some letters. When Abu Beker came to get me to go to the steamer he had his little girl with him, a child about 6 years old. I was surprised to see her ornamented like a Kling woman with silver anklets, four holes pierced in the top of each ear each with a silver ring in it, and a hole in each lobe. One of her nostrils was also pierced and had a gold ring in it and her finger nails were dyed red. He is an Indian from the Bombay Presidency and follows the customs he learned from his parents. We went to the mosque for prayers at 1 o'clock and there, for the first time, a sort of protest against my dress was made by an old man who looked daggers at me and told me I ought to have different pants on. He proved to be a beggar who had come in from the mountains and was considered unevenly balanced mentally. I was advised to pay no attention to him. Generally I was treated with extreme courtesy. Returning we saw the method of distributing letters to the various races, each having a separate place on the street. Smoked the hookah for the first time and found it rather pleasant. Came home in the rain about 4 and as it rained all the evening there was nothing to do but sit in the house and chat. Went to bed at 9.

October 2, 1892

Up at 4. Rained all night and cleared a little about 8 a.m. Bro Abdulla Arab went to town about 7:30 and left me to write and enjoy myself in my own way. Very grateful for the relief as I get very little time to myself now. There is always someone about to take up my time and sometimes I am excessively bored. While I was taking a comfortable nap in came Abu Beker and a young man of European inclinations and dress and invited me to visit the Shoay Dagon or Golden

99

Pagoda. I was, of course, very glad to go and we started off
in a gharry. We first visited a native market where all sorts of
curious and indescribable eatables were sold by Burmese and
Indian men and women—rotten fish made into a muddy
paste—leaves, herbs and grains—ear-rings and trinkets—
Indian drug shop with roots and herbs. After this we started
for the Pagoda passing a number of Buddhist temples which
I was very anxious to see, but Abu Beker had a weakness for
Sesamum oil mills. The Golden Pagoda—beggars at entrance and huge white dragons of brick and mortar—curious
stucco-work figures over arched entrance to first stairway,—
peddlers of flowers, toys, trinkets, pictures, sweets, jewelry,
candles and altar offerings. Several flights of stairs with curious stucco ornaments all the way up. Came to level court in
which the famous pagoda or monument stood. Said to be
hollow,—two Englishmen found secret entrance once and
only one came out alive—Buddhists said spirits killed him—
mistake—disappointed at not finding whisky and soda in
there and broke his heart. Gorgeous vista—flag on top
looked a foot long—actually 10—elephants couchant and
niches with many Buddhas around base. Company of very
clever native opera bouffe artists in one of the temples. Blind
fiddler, cock-eyed pipe player, second fiddler and two
women squatted on pavement amusing large crowd. 1st fiddler very funny duo with one of the women,—piper acting as
clown—enjoyed performance very much but Abu Beker was
impatient and wanted to go—hadn't seen oil mill for nearly
half an hour. Multitudes of Buddhas of gilt, marble, silver
and bronze in cages gorgeously decorated. Musicians and
fortune tellers, and peddlers of food. Regular circus of gilt
and tinsel. Blind beggars innumerable one with a hole in his
face—horrible. Beautiful view of the city and surrounding
country from the southeast wall. Regretted that I had not
more time in which to inspect the various temples. In the afternoon about 4 went to the down-town room and at 7 had
dinner with the sick man. The Imam, Sufi and several very
prominent Mussulmans present. Squatted on the floor and ate
in the Mohammedan style. Getting rather accustomed to it
now. Was introduced to the leading merchant of Rangoon—a
Moslem—who expressed the deepest interest in the Ameri-

can Propaganda. Abu Beker accompanied me home—too late to see oil mills but he knew of one, he said, that we could go to see to-morrow. Abdulla Arab remained in town all night.

October 3, 1892

Visited the Post Office and mailed letters to wife and Kelsoe, and then went to the principal bazaar of the town and would have enjoyed it very much if I had not been surrounded and followed constantly by curious Mohammedans who were anxious to see the American convert.[196] Was repeatedly invited to sit down in the shops and attract trade. Saw many curious things. The Bazaar is a vast affair and covers perhaps three ordinary blocks. It is occupied by retail dealers in everything used or worn from the humble corkscrew to the silk-dress. In the vegetable department there was a fine display of most luscious looking fruits and vegetables, but our brethren do not seem to use them. My mouth watered as I saw the delicious antis and the guava,—the finest I ever saw—the yams and sweet potatoes, and cucumbers and crisp lettuce. My table has been supplied only with meat, meat, meat,—seasoned beyond my taste and heavy with grease. Oh, for a good fruit and vegetable dinner, for a change. Richly embroidered caps, turban cloth, silk hdkfs. &c. Burmese brass ware, drugs, jewelry and notions mingled with fancy goods from Europe. The most toothsome candies I ever saw—pure white and richly flavored. Sweets of various kinds in bowls with crystal syrup. On the way to Abu Beker's shop we stopped at a little shop in a squalid side street and saw some most artistic specimens of carving in ivory done by the Burmese. One

---

[196] William A. Kelsoe, 1851-1932, a longtime friend of Webb who was city editor at the *St. Louis Republic* as well a councilor for the St. Louis branch of the Theosophical Society. Walter B. Stevens, *100 Years of the St. Louis Republic*, 21; Hyde and Conrad, *Encyclopedia of the History of St. Louis*, 4:226. Some of the letters Webb sent to Kelsoe on this trip were published in the *Republic*—"The Land of the Pagodas," *St. Louis Republic*, December 4, 1892, p. 10; and "Looking for a Yogi," *St. Louis Republic*, December 12, 1892, p. 3.

workman was engaged in carving a large elephant's tusk, and another was at work on a smaller piece of ivory. Both used pieces of deer horns to hammer their steel tools with. A paper knife about 18 inches long with a beautifully carved handle and engraved silver ferule was offered for 25 R's or about $11 Mexican dollars, or $7.85 U.S. currency.[197] Naked children with silver shield hanging over privates, silver anklets and bracelets, earrings toe rings and finger rings. Had a drive past the race-course—gymkhana, and many very attractive residences and private grounds. Visited also Mohammedan school of 1st and 2nd classes—about 50 children taught by 2 teachers—greeted me with the chorus of "Salaam Aleikum."[198] Came home about 10:30 and had breakfast with Bro. Abdulla Arab and my escort after which they went down town and left me to enjoy a nap. As usual the pestiferous, useless boys came tramping up stairs and woke me up. Of all the nasty, filthy, worthless beasts I ever saw the black boy who seems to have charge of the house is the most. I think his stay would be terminated very abruptly if it was my house, but as it isn't I must submit. After Zohar took the gharry and drove down the Kemendine Road to some sheds where four elephants were kept. These animals are used in the saw-mills and lumber yards to move heavy pieces of timber. The 4 immense creatures,—2 of them apparently as large as Cole's "Sampson" or "Jumbo" were quietly fanning themselves with their great ears and eating cut grass which was strewn on the floor around them.[199] They were owned

---

[197] Mexican dollars were silver coins minted in Mexico that were widely circulated in Southeast Asia, including British colonies. P.J. Drake, "Southeast Asian Monies and the Problem of a Common Measure," 92-94.

[198] The school was most likely the Madrassa Islamia. "News Notes," *Moslem World and Voice of Islam*, November, 1895, p. 3. "Salaam Aleikum" is the universal Muslim greeting and translates from Arabic as "Peace be upon you."

[199] Sampson and Jumbo was a reference to the "Elephant Wars" of the early 1880s whereby the major circus companies tried to outdo each other through owning the world's largest elephant. William Washington Cole, 1847-1915, purchased "Samson" (Webb refers to it as "Sampson") in response to P.T. Barnum's, 1810-1891, acquisition of

by native Burmese who hire them out to the saw-mills by the day. Then went to another mill where an elephant was at work and witnessed one of the most wonderful performances I ever saw. The animal was perfectly trained and seemed to understand everything his Mahout, or driver, said to him. He moved about with his trunk, tusks and feet, heavy pieces of wet timber and looks that it would require 20 men to lift, placed them exactly in the position necessary; selected other pieces and carried them away to their proper piles and acted with just as much intelligence and judgment as an ordinary workman. Drove to Abu Beker's shop and then to the down-town room where, after Asr, we drove with Bro. Abdulla Arab, through some of the principal residence streets and out to Dalhousie Park, the most attractive spot, I am told, in Burma.[200] In it is situated the Royal Lake from which the city's water supply is drawn—it is rather a chain of lakes of rather muddy water at this time of the year when it rains the greater portion of the time. In the dry season it must be very clear. The park is most beautifully laid out and the closely cropped lawns dotted with great shade trees present a very attractive vista. There are neat little pagodas and rest houses nestling among flowers and on one of the larger lakes are a number of skiffs or row-boats for hire. From the highest point of land, on which a pretty pagoda is situated the view is superb. Little islands with tropical vegetation and inviting coves and velvety green promontories. The drive home was through wooded glens and past finely laid out grounds, and although a drizzling rain had set in and the roads were sticky, it was very pleasant.

"Jumbo." William L. Slout, *Chilly Billy: The Evolution of a Circus Millionaire*, 189-91.

[200] Dalhousie Park was named after James Andrew Broun Ramsay, 1st Earl of Dalhousie, 1812-1860, who was governor-general of India and lead the invasion of Burma in 1852. *Encyclopaedia Britannica*, 11th ed., s.v. "Dalhousie, James Andrew Broun Ramsay."

October 4, 1892

For the first time in several days there was no rain this morning and it was clear and pleasant all day—the first wholly clear day we have had since I have been here. In the forenoon arranged draft at bank and in the afternoon went to some saw-mills owned by Mohammedans and saw some elephants at work. One log that it required 6 men and a hanging lever or derrick to raise to its place on the saw bench, the elephant lifted and dragged along with comparative ease. The strength and intelligence displayed by these brutes is something marvelous beyond anything I ever saw. Was accompanied by Abu Beker who seems to have recovered from his oil-mill fever and is attacked by rice-mills. On the way back encountered a Buddhist religious procession which was very interesting. At this time of the year the Buddhists here make contributions for the support of their monasteries giving the monks all sorts of things from dishes, brooms, mats and other household articles to fruits and every sort of food. What the name or explanation of this feast is I have been unable to ascertain as my Mussulman friends seem to know nothing of the matter and they do not seem inclined to allow me to visit the Buddhist temples. These contributions are carried through the streets accompanied by gaily dressed girls and boys, and a bullock cart with Burmese musicians in it. Their music consists of tom-toms, cymbals and pipes or flageolets. In front of the procession boys and girls skip and dance along twisting their bodies into various contortions. A grotesque figure of a woman about ten feet high made of bamboo covered with painted paper and with a man inside to animate it, gambolled clumsily along in the rear in company with other grotesque figures. In another street we saw collections of food and household goods being made from door to door almost every house contributing something. The streets were filled with gaily dressed Burmese Buddhists and all seemed to enter heartily into the spirit of the season. In the evening a large meeting of Mussulmans was held in the Mohammedan school hall opposite the mosque which I addressed at considerable length on the subject of the American mission and was listened to most attentively. At the conclusion of my address the Imam of the mosque made a very eloquent speech in Arabic endorsing the project. Quite a large number of subscriptions were made and more are expected to-morrow. Brother Abdulla Arab and myself reached home shortly before 11.

October 5, 1892

Before daybreak the notes of pipes and tom-toms came to us from one of the neighboring Burmese villages and unless they have been keeping it up all night the festivities of the day have commenced. To-day and to-morrow are festival days and business is suspended. After desayuno started out to have another look at the Golden Pagoda, hoping to see the blind musicians again and have a more leisurely view of the temples. It was evidently a prayer day and no musicians were there. But in their place were numerous yellow-robed priests with shaved heads, squatted on the ground ringing bells or calling aloud for alms; some of them appeared to be doing a very good business. There were peddlers innumerable, many Chinamen and some Mohammedans and Hindus showing that the Buddhists here are tolerant. Councils of dirty-faced gods—hot-headed gods cooled by throwing cups of water over them—great bells weighing many tons—and smaller ones very elaborately ornamented—wonderful wood-carving about the temples—one had fine trellis work of vines carved out of solid teak a most exquisite and artistic piece of work. Family parties squatted in temples eating—some temples in very bad state—need repairs. Offerings to the spirits of the Pagoda eaten by flocks of clamorous crows. But what an imposing structure the pagoda is! at least 100 feet at its base and towering up to a dizzy height. Workmen ¼ way up cleaning, look like pigmies. Chinaman making toy figures of monkeys, rats, fruit &c "while you wait" for 1 Pice. On the way down the steps bought a Burmese pipe and two song books—Burmese just like all other thieving shop-keepers—first man demanded 4 Annas for book, next 8, next 2 Steam tram down to China St for 6 Annas,—distance about 1 mile. Took gharry home—legal fare 4 Annas—Syce demanded 1 Rupee and insisted on having it. He got 7 Annas, 3 more than he was entitled to and went away growling. Put in the afternoon in reading, sleeping and revising my lecture for the following evening. Went down town from 3 to 4 to purchase Burmese clogs for bath but found Burmese all gone to pagodas. Bought some cheroots as feared would not get any—in evening Abdul Kareem redeemed his promise made several days ago and bought me some. Promises easily made and soon broken. Saw Chinese theatre down town on street.

October 6, 1892

Morning clear and pleasant. and weather remained fine all day. After breakfast went down town to Abdul Careem's shop and saw a Burmese funeral made up and started. But I should say first that about 8:30 I heard a man singing at the top of his voice and looking out of the window I saw an enormous elephant striding up the road swinging his trunk and flapping his ears as if he was enjoying himself while the mahout sat on his neck and sang merrily a Burmese song. He passed up the road and the notes of the song died away in the distance. Presently I heard the song again and it was drawing nearer; soon the first elephant appeared in sight with two others following each with a mahout on its neck. They swung lazily along, reaching out with their trunks occasionally by the way and occasionally stopping to eat while the mahout chatted. One of the animals had a sugar cane in his trunk from which he bit a piece and crunched noisily just as a boy would. They seemed perfectly happy and contented—both animals and mahouts. The Burmese funeral was a curious and imposing affair. The body had been dead a week but as it was in an air tight box it offended no one. The catafalque on which box rested gorgeous with colored paper and gold and silver tinsel as was an immense oblong canopy which was borne over it by ten or 15 men. The bed on which died ornamented with colored cloth and two Turkish rugs on it—to be given to monks with about a dozen bullock cart loads of fire-wood, rice, fruit and provision of various sorts, brooms and towels. It is the custom on the occasion of the burial of a Burmese to present to the monks as great a quantity of provisions and household articles as the relatives can

106

afford to pay for. The greater the value and quantity of stuff given to the priests or monks the better will be the chances of the deceased's selection for salvation and the sooner will the soul of the deceased go to nirvana. The frequent funerals must stock the monasteries with an immense supply of household goods, and it was said to me that they usually sold the greater portion of them at very cheap rates. Where the procession started a Burmese band with tom-toms, cymbals &c headed it and the bullock-carts full of provisions &c followed, the corpse brining up the rear. At intervals between the carts were Burmese bands, the players all decked out in their brightest colors and presenting with the gorgeous catafalque and canopy, a very brilliant scene. In the afternoon at 5 dined with a number of the leading Mussulmans at the country residence of Ismail Mohammed who sent his elegant carriage with driver, tiger and two footmen to take us to his house. The carriage cost, I was told 1,800 rupees in Calcutta. Ismail is a young man whose father died some time ago and left him a fortune. He lives in fine style in a garden filled with tropical trees and flowers and furnished the best dinner, by far, that I have had here, or anywhere in the East. The older Moslems squatted on the floor, while a table with bran-new silver ware was furnished for Abdulla, myself and two others who spoke English. Handsome house embowered in trees and vines. After dinner we had prayers under an immense tree in the garden, the Sufi moulvi leading, and then drove to the hall where I delivered my first lecture on "The Faith of Islam."[201] By God's help it was a success so far as I was concerned and I was surprised at myself. As a lecturer I'll do. There were only about a dozen Englishmen present, three ladies, some Parsees, and Eurasians and the rest of the hall was filled with Mohammedans who could not understand English. At the close of my lecture a Parsee who seemed to be offended because I had not alluded to his religion and who seemed to be pretty well stirred up over something. Finally he apologized for having spoken to me and

---

[201] The only surviving text of this speech is an Urdu translation in the work *Ishaat i Islam* (1893), available at the Oriental and India Office Collections of the British Library.

went away. I'm glad he did for I am afraid I would have lost my patience as he was an Anglo-maniac. We reached home at 9.

October 7, 1892

Another nice morning—about daybreak took a long walk in the garden surrounding the house—beautiful place filled with great shade trees, shrubbery and flowers. About 10 went down town and arranged about passage to Calcutta—found the usual insolent, conceited English cad to deal with at the agent's office. It does seem strange that Englishmen cannot be gentlemen. Saw nothing of special interest during the forenoon and went to the mosque for prayers at 12,—mosque filled to doors and some praying outside. Hindu too lazy to sit up and knit his net, lying on the door-step to do it. Sun excessively hot. Paper failed to notice my lecture—poor sheets—no local news. *Gazetta* and *Times*—probably run by two boys and a man. 10 pages of ads and four of very tame stale reading matter,—clippings. One editorial a day which was all the editor's English nob could stand. In the morning saw 3 elephants in front of the house eating leaves from the trees and apparently enjoying themselves while their mahouts lounged lazily on their backs. Abu Beker wanted to visit an oil-mill and spoke of it several times but as it was too hot and I manifested no interest in the project he decided to postpone it until to-morrow morning. That settles it—he'll forget it. In the afternoon he and Ahmed Moolla Dawood and myself visited the public museum and zoological garden, where we found a fairly good collection of animate and inanimate curiosities badly kept.[202] The garden is badly laid

---

[202] Ahmed Moolla Dawood was a partner in the merchant, trade, and rice mill firm of Moola Dawood, Sons & Co. He contributed $300 for Webb's cause in 1894 and provided literature and translations of articles on Islamic topics for Webb to publish. *Cyclopedia of India*, 3:471; "Mohammed Webb's Account," *New York Times*, March 27, 1896, p. 3; "News Notes," *Moslem World and Voice of Islam*, August, 1895, p. 3.; and "Personal," *Moslem World*, November 1893, p. 9.

out and has a very slovenly and neglected appearance. The animals are caged in a cheap rude manner and the cages and surroundings of the feline specimens not being properly cleaned emit a stench that is overpowering and drives visitors away. The collection of monkeys is the largest and most complete I have ever seen and the white elephant given by, or stolen from, the king of the country is a good attraction. He has been trained to bow and trumpet when visitors are around to get apples, sugar cane fruit &c. Sugar cane is his weakness. A white monkey with the pink, restless eyes of the albino and having the face and manner of a palzied old man is one of the rare attractions as well as a Burmese black tiger. Under proper management the zoological collection could be made very attractive, but the poor, shabby cages and thin bad arrangement spoils the whole show. There is a fairly good collection of stuffed birds in the museum and a very good exhibition of Burmese and Indian seeds and medicinal plants. None of the animal cages, of course, are labeled and visitors have to guess.

October 8, 1892

Another clear and bright day. We are to leave for Calcutta at 4 to-morrow morning on the S.S. "Nerbudda" of the British India Navigation Co's line and are to go on board this evening about 8. Nothing of importance occurred during the forenoon. On the way down-town stopped at sale room of jail—small assortment of ... made by prisoners. Some fine carving. In the afternoon Abu Beker succeeded in getting me to an oil-mill—a small badly arranged petroleum refinery. An Englishman showed us around and seemed to be rather ashamed of the place—but of course—they could produce better petroleum than America but did not want to. That stupid asinine English conceit will crop out. The whole concern was a sort of parody on a first-class petroleum refinery—American and Russian boxes and tins planed and repaired and passed off as new. Went to mosque at 4 and then home where finished packing and trunks were sent down to steamer. We followed later and went direct to mosque. After prayers I was called upon to say good-bye and made a brief speech which was translated by Abdul Karim and responded to by the Sufi. Abu Beker then made a brief farewell address to which I responded briefly. Then the hand shaking began and it was with great difficulty that I made my way through the crowd to the door. We then walked to the steamer where I was again besieged by enthusiastic brethren who wanted to shake my hand for the last time. Went to bed at 9:45 leaving Abdulla Arab talking to one of the visitors.

October 9, 1892

Arose at 4:15 just as steamer was leaving the wharf. Bright, clear day. Young Briton on deck at 5 with jag on. Very full. By 8:15 we had reached the end of Elephant Point, discharged the pilot and turned to the right toward the Bay of Bengal. Curious effect on sea opposite the mouths of the Irawaddy—patches of muddy water, color of clay and patches of deep blue alternating. Proved to be cloud shadows. It was after noon before we came into clear water steaming at the rate of about 14 miles an hour. The mud from the Irawaddy must discolor the water for at least 50 miles around its mouths. Our fellow passengers seem to be a quiet set. The young man with the jag showed up later in the day in very fair shape but looking somewhat the worse for wear. There are seven men and one lady—a Mrs. Wm. Thompson who is accompanied by her husband and two wooden reclining chairs with the name Wm. Thompson "in large" black letters on the backs so that he who views may read and know that Thompson, with a P, is on board. Mrs. T. is rather old and worn looking while he looks as if he had dissipated a good deal. All the men are young—from 25 to 35—except one and they look as if they might be clerks or army attaches of some sort. One has his head shaved close a round head and a well-kept blond mustache. He is apparently given to the flowing bowl and women. He has a rakish, sporting look. All the men, with one or two exceptions are afflicted with the habit of coming on deck bare-foot just after it has been washed and prancing up and down for half an hour or more as if they were walking for wages. There seems no good reason why they should do this except that it is a fad which some English tourist has started and the rest of his country-men are in duty bound to follow it, whenever they go to sea, until some beefy duke, or lord or marquis or some other ti-tled ass invents a new piece of idiocy to take its place. We were in the Bay or Sea of Bengal before noon and all day it behaved itself most decently. This body of water has a bad reputation for tantrums and makes itself felt at nearly all sea-sons of the year; but to-day it is on its good behavior and is as quiet as a mill-pond. There is just breeze enough to make the sail pleasant and there is little or no motion to the vessel. All meals were served on deck. Sitting aft reading in the af-ternoon I saw a little bird come out from under the rudder box and thought it had escaped from a cage somewhere on board. But it proved to be of the wren family and had been carried out to sea in some unaccountable way. It hopped about the deck quite fearlessly and then went forward on the awning. I saw it about the ship later in the day and it is probably stowed away somewhere.

October 10, 1892

Morning broke beautifully with a fine stiff land breeze which did not give the vessel any motion. Nothing of interest but several schools of flying fish—little bird remained with the ship flying about all day and chirping merrily. Brisk shore breeze all day but sea comparatively quiet—very little mo-tion. Passengers amused themselves with cards—Abdulla Arab and I slept and read.

IV.

<u>CALCUTTA</u>

<u>October 11, 1892</u>

Another beautiful day with just a slight ripple on the sea—wonderful weather for the Bay of Bengal—the voyage as quiet and enjoyable as a sail up the Hudson. Saw the sun rise from a clear sky and smooth sea—beautiful effect—1st pink, yellow, crimson, bright gold and ball of fire emerged slowly from water. About noon we came in sight of the first pilot ship and an hour afterward passed another with the word "Intermediate" painted in large white letters nearly the full length of its broadside. Half an hour afterward we saw another pilot ship marked "Lower Casper" and soon afterward another marked "Upper Casper". We passed a sailing vessel, a steamer and a tow boat, but no land came in sight until about 3 when we could see off to the right a long low black strip which we were told was India. We were nearing the mouth of the Hooghly. Had pilot with us brought from Rangoon. Sea gulls joined us about 3 and acted differently from Singapore, Penang or Rangoon gulls—former tipped the water delicately with their bills—Calcutta ones plunged in and seized fish. About 4:15 we passed another pilot ship with the words "Long Sand" on its side and in the distance to the right appeared a light house on a flat tongue of land surrounded by small buildings. The land at the right appeared plainer but very low. Water began to get dirty clay color about 2. About 5 it changed to a light brown—we had passed the light-house and a long dreary stretch of low flat land spread out on the right while on the left no land at all was visible. There were no hills to be seen. At 5:15 we dropped anchor apparently 3 miles from this bleak shore and with not a house or a vessel in sight. The tide was too low to admit of our entering the river. As there was no breeze caused by the movement of the vessel the heat was felt and most of the male passengers slept on deck—my cabin however was very comfortable, a light land breeze flowing in at the window so strong that I had to close the door.

October 12, 1892

Flocks of devil's darning needles came off to see us as we lay at anchor—thousand of them—land in long low strips faintly seen on the left.

In her novel *Felix Holt* George Eliot makes the hero say to Esther Lyon:

> "I can't bear to see you going the way of the foolish women who spoil men's lives. Men can't help loving them and as they make themselves slaves to the petty desires of petty creatures. That's the way those who might do better spend their lives for nought—get checked in every great effort—toil with brain and limb for things that have no more to do with a manly life than tarts and confectionary. That's what makes women a curse; all life is stunted to suit their littleness."[203]

[And again:]

> "Truth is the precious harvest of the earth
> But once when harvest waved upon a land,
> The noisome cankerworm and caterpillar,
> Locusts, and all the swarming foul-born broods,

---

[203] *Felix Holt, The Radical* (1863). Quote from chapter 10.

113

Fastened upon it with swift, greedy jaws,
And turned the harvest into pestilence
Until men said: 'What profits it to sow?'"[204]

At 7:20 we raised anchor and resumed our course the water having risen. We were in the Hooghly River which, at its mouth seems to be 10 miles wide, at least. The shore as we steamed up was flat, and covered with a thick jungle growth, with no signs of habitation save an occasional light-house. Englishman said that west shore was infested with tigers and very dangerous to land there; but since an Englishman told me that all the newest inventions of the age were emanations from England and that that country was the greatest in the world I have become skeptical. Crying Indian bay on board—deck—cries the most melancholy—sends cold chill all over me and makes me think of funerals. Shortly before 10 signs of life began to appear along the left shore—grass-thatched huts and bungalows nestling among the trees—droves of cows grazing—natives fishing with nets along the muddy shore which crumbled and washed away under the action of the swell from passing steamers—native boats &c. Stream and banks resembled those of our Mississippi's clay color and mud. Two steamers the "Ooryia" and "Sea Gull" the latter an iron side-wheeler passed us at 10:10 bound for the Indian coast ports. And this is India at last—am I dreaming? Ebony natives on shore naked—but for white cloth around loins and bare-headed in the broiling sun pulling a boat. At 11:30 we stopped off "Diamond Harbor" a railway station with two or three stone bungalows and a lot of fisherman's huts around it, and were boarded by a native with letters and telegrams for the ship. Sometimes mails sent from this point to Calcutta when steamer is late or tide too low. Stern-wheel steamers with 2 smoke-stacks forward like old Mississippi boats. Garden Reach and the Palace of the King of Oudh.[205] About 3 arrived opposite Calcutta and turned

---

[204] Ibid., quote begins chapter 11.

[205] Wajid Ali Shah, 1827-1887, was the last independent sovereign of Oudh, when the British annexed his territory and exiled he and his

around to back up to wharf. Many sailing vessels in river—signs of extensive traffic. It was nearly 4:30 when we arrived opposite our landing place and found that the water was too low to admit of our going in. Several Mussulmans came off in a sampan and escorted us and our baggage ashore where we were received by about 1,000 Mohammedans who gave me even a more cordial reception than I had in Rangoon. We escorted to the house of Hajee Noor Mohammed Jackeriah, where we were given very pleasant quarters and where I was greeted by a large number of Mohammedans.[206] Mosquitos too bad to sleep.

October 13, 1892

Mosque before daylight. The mosquitos having destroyed my rest the previous night I was a total wreck and passed most of the morning in sleep on a lounge in the parlor. About 2 received a call from the Nawab Abdool Latif who was formerly a magistrate but is now living on a government pension and is considered a very learned and very good man.[207] I found him a little inclined to be stiff at first but he

---

family to Calcutta in 1856. Parshotam Mehra, *A Dictionary of Modern Indian History, 1707-1947*, 790.

[206] Hajee Noor Mohammed was vice-president of the Central National Mahommedan Association. Also, possibly Hajee Nour Mohammed Abu Talib who was a member of the committee to raise funds for Webb's mission. Muhammad Y. Abbasi, comp., *Annals of the Central National Mahommedan Association, 1878-1888*, 12; and Webb, *Islam in America*, 69. His home was located in Amratolla streeet. "Current Events," *Times of India*, October 20, 1892, p. 5.

[207] Nawab Abdul Latif, 1828-1893, spent most of his career in judicial service to the government, including Presidency Magistrate, Calcutta. He was also influential in promoting Muslim education and founded the Mohammedan Literary Society in 1882. Upon the Nawab's death Webb wrote diplomatically of him, stating: "Whatever may have been his faults or his virtues he was certainly a most companionable man; an entertaining, interesting conversationalist and a genial host." "A Useful Career Ended," *The Moslem World*, September 1893, p. 10; and Ahmad Saeed, *Muslim India (1857-1947): A Biographical Dictionary*, 21-22.

soon unbent and became quite talkative. He is not a fool and his success is apparently due to his slippery but self-contained, calm methods. His eye is always on the main chance and he is very clever at hedging. If he makes a mistake he slips out of it very neatly and makes it appear intentional. He knows who to fawn to and when. Altogether he is a harmless good-natured old fellow and I enjoyed my talk with him. At 4:20 Hadjee Noor Mohammed took me to the mosque in his elegant carriage, with the two sikh footmen yelling at the passers (a curious custom) and presenting a very dignified appearance. After prayers we drove through some of the principal streets and to the Maidan along the bank of the river opposite the shipping. Saw the Viceroy's house and grounds—very pretty—lion over gates with his tail rampant. Imposing statues of English army officers who have distinguished themselves, Lord Ripon et al.[208] Drives very pretty, the maidan laid out in smooth lawns and the inevitable athletic ground. Passed a public garden and caught glimpses of green vistas and cool, shady nooks that seemed very inviting. The general aspect of the parts of the city through which we drove gave me the idea that Calcutta is a very handsome, well-kept city. Viceroy received 25,000 Rupees a month, lives in Simla from March to December and has little or nothing to do.[209] The natives pay the freight. After dinner at 7 went out for a walk through the streets accompanied by Hajee Abdulla's nephew and the usual escort. When a few yards from the house a Moslem who could speak English came up and assumed the position of guide. His unasked services were accepted gladly—it was the best I could do but I would very much have preferred to be left alone. Why can't these people show me something that I want to see; they will persist in dragging me about to places

---

[208] George Frederick Samuel Ripon, first Marquis of Ripon, 1827-1909, Viceroy (1880-1884). Know mainly for his reform policies and his failed attempts to cede local control to the Indians. Mehra, *Dictionary of Modern Indian History*, 611-14.

[209] The Viceroy in 1892 was Henry Charles Keith Petty-Fitzmaurice, fifth Marquis of Lansdowne, 1845-1927. He held this position from 1888-1894. Ibid., 388-91.

that I don't care a rap for and carefully avoid the very things I want to see. On this occasion they dragged me off to see an old pontoon bridge that hadn't the slightest interest to me and could only be seen partially in the dark. Caught a glimpse of a Hindu bathing ghat which I expect to see later on and of the interior of a Hindu temple with big-eyed gods and doll gods. Also saw a man with testicles enlarged so that they reached to his ankles and interfered with his walking. Curious disease said to result from drinking cocoanut milk at a certain stage of growth. Native shops always full of interest. Came home about 9:30 and Hajee Abdulla invited me to go to a Mohammedan wedding but I was too tired and, as there was to be another in two days, I went to bed. Bed not arranged,—promises.

October 14, 1892

Slept later than usual and did not get to mosque until after prayers had been said. Returned and had milk, oranges, papayas and toast. Called on Gen. Merrill the American Consul General and had a pleasant chat with him—rather stiff at first but soon melted.[210] At noon after prayers at mosque addressed a large gathering of Mohammedans there—estimated at 5,000. Translation made by prominent Mohammedan lawyer. After the close of my lecture the people crowded around me trying to get a chance to shake my hand, and it was with the utmost difficulty and the active help of several Mohammedans who joined hands and formed a circle around me, that I was saved from being crushed. I was led to the upper portion of one of the adjoining buildings and the lower doors were closed so that no one could reach me. The people seemed overjoyed to see me and wild to get at my hand. It showed how deep and enthusiastic is the interest manifested in the American mission. Spent the afternoon in a chat with several intelligent English-speaking Mohammedans and re-

---

[210] Samuel Merrill, 1831-1924, native of Indianapolis, Ind. Began his career as a publisher and bookseller. Served as Consul-General at Calcutta from 1889-1893. *Who Was Who in America*, 1: 832.

ceived an invitation to lecture at Madrassah Hall.[211] After dinner drove out intending to visit the Eden Garden where a band plays three times a week, but found it two late—the electric lights had been turned off. We drove about the city, saw the large market and many native shops by lamp-light. There was nothing of importance to record except that the streets were strongly Anglo-Indian and their oriental appearance was spoiled by the presence of numerous toddy and beer shops and general English proclivities. When I came home bright native man questioned me—keeps a book store, sells old stamps and clerks in an office. "Does queen Victoria rule your country? We have a Greek Consul here; are you a Greek? What are you? American? Oh! Where do you live? In America? Oh! What is you name in America? Care of who? What telegraph address? You don't live in Manila? Manila is not in America?

October 15, 1892

After prayers at the mosque went with Hajee Abdulla and his nephew to see some Hindu ghats—bathing in the sacred river,—that is, the Ganges joins the Hooghly above and that makes the later sacred. Hindus, male and female smeared with white and red marks—ascetics covered with flour or dust—one wild-eyed man with hair twisted into ropes and face covered with dust. Others pouring water over little piles of stones, black boulders, and watering trees—offering flowers to the stones and trees and pouring water over them. Strange sight—visited three ghats—the tide low and water muddy. Went then to the Eden Garden,—beautiful place lake and water lilies—palms—lawns &c. About 10 went to the Chartered Bank of India, Australia and China,—repetition of the experience in Singapore—1st clerk didn't know where draft dept was—native came to his rescue and suggested that I had better go inside an enclosure and speak to a man at desk—man didn't know whether he could sell draft on Frisco

---

[211] See discussion below concerning Madrassah College on October 16, 1892.

or not but would ask[212]—could—native escorted me to waiting room—sat there ½ hour—much discussion and running
about—took me into another room where there was a smiling
half-caste saying "that man" would give me draft and
money—"that man"—proceeded to ask me where I came
from, when going, and whether ever coming back to India—
answers seemed to please him. After deal of figuring gave
me draft and pointed to several men behind desk on opposite
side of the room saying they would give me the bal of 360
Rs 5 A's, "But" I said they don't know me and I have nothing to show that I am entitled to 360 R's—would go, he said,
and mention matter to them—did so and they all looked at
me in blank amazement. Waited until I got tired and went
back to the smiling man telling him I saw no prospect of my
getting money—he went to beefy Englishman who delivered
up old draft and smiles took it and went to cash men. They
regarded it with great curiosity, discussed it for some mo-
ments—wrote something on the back of it and discussed it
again. Then they concluded it might be safe to pay me the
360 R's but I must go to another desk to get the 5 A's. A little after 4 went out riding with rapid half-caste in dog cart—
racing horse—reckless driving—collided with bullock cart—
knocked down one of the bullocks—nothing broke but pur-
sued our mad career having half a dozen very narrow escapes
from smash-up—said prayers on Maidan and listened to Eng
military band at Eden Garden. Englishmen are not musicians. Got home alive. In evening after 8 went to Mohammedan residence to see celebration of festival of Mowlood
Sheriff or birth of Mohammed. Fully 500 people squatted on
floor hundreds of candles in crystal chandeliers and colored
globes—old man seated on floor reciting and commenting on
Koran. When concluded, sweetened, spiced milk and short-
cake served,—very delicious. I was given seat of honor at
the right of the reciter and left of Dewan Abdool Latif. Tried
sitting cross legged—introduced to the Prince of Oudh.[213]

---

[212] Presumably "Frisco" refers to San Francisco.

[213] Kamar Qadr Mirza Mohammad Abid Ali, 1852-1919, the eldest son
of the last King of Oudh and head of the family from 1888-1919.
Who's Who in India, 5:97.

Said wanted to go to America but Gov't wouldn't let him—wanted to complete medical studies.

October 16, 1892

Intended to go to Botanical Garden but remained at home to receive a call from Abdul Jubbar and Nawab Syud Ameer Hossein the latter a Justice of the Police Court; but they did not call until 5 p.m. much to my disappointment.[214] They are quite intelligent and good-looking men. Spent the day in preparing my lecture to be delivered at night before Madrissa Debating Club.[215] The hall of the Madrissa College had been promised by the headmaster who, late in the day, changed his mind, saying that as I was an American and might have something about politics in my address, it would not do to allow me to speak in the college hall. I suspect that the real reason was that missionary influence was brought to bear on him. The missionary and the Britishes are all powerful here and there seems to be a general disposition among the natives to cringe to them. A smaller hall, and one not altogether comfortable, was secured and my lecture was received with much enthusiasm. The usual vote of thanks was passed and I was assured that I had greatly entertained the gathering.

---

[214] Moulvi Abdul Jubbar, Khan Bahadur, 1836-1918. A founding member of Anjuman-i-Islam and an early supporter of the militant Wahabi movement in India. He was later the president of the Central National Mohammedan Association. Jayanti Maitra, *Muslim Politics in Bengal, 1855-1906*, 23, 75; and Saeed, *Muslim India (1857-1947)*, 18. Nawab Syed Ameer Hossein, ?-1910, held several high government positions, including Inspector General of Registration (1893-94) in Calcutta and was secretary of the Central National Mohammedan Association for 18 years. Ibid., 81.

[215] The Madrassah Debating Club had been part of the Calcutta Madrassah since at least 1870 and predated the founding of Madrassah College, which began in earnest in 1883-84. Abbasi, *Annals of the Central National Mahommedan Association*, 80.

October 17, 1892

After prayers this morning I was accompanied to the Zoological Garden by Jan Mahomed Gangjee, of 63 Ezra Street, who used his own gharry.[216] The zoological collection is the most complete and best arranged I have seen outside of America. The only drawbacks were that some of the cages were not kept clean and very few of them were labelled so that so that the visitor could know what animals were inside. But of course one should not expect too much from the British way of doing things. Our couzins are slow and it takes a long time for them to get anywhere—they cannot keep up with the times—too fast for them. The reptile house is the finest of its kind I ever saw and the snakes are shown to the best possible advantage. It is made of brick and marble in is a most attractive structure. The collections of birds and monkeys are first class and there are a number of rare land and water animals—they have 3 very fine Brazilian tapirs. The grounds are beautifully laid out and remind me very much of the St. Louis Zoological garden. The cages will average fully as good as those at St. Louis. We hurried through and did not have as satisfactory a view as I would have liked as Bro. Hajee Abdulla had made an engagement to call on Abdool Latif and wanted us to return at 8 or 8:30. We returned at 9 to find that the call had been postponed until the following day. We might spent the whole forenoon at the garden. The collection of monkeys is quite as good as that at Rangoon, if not better, and there are some specimens that I never saw before. There are several fine specimens of the Hanuman or sacred monkey, and are called the Hoo-loo who calls out "Hoo-loo" in a voice that can be heard half a mile, at the slightest provocation. The lions and tigers are also good specimens and well kept. Went to Post Office about 10 and sent registered letter to Ella containing draft for $250.00. In the afternoon received letter from H. Dharmapala telling me that Col. Olcutt

---

[216] Jan Mohammed Ganjee was later Webb's agent in Calcutta for his publication *The Moslem World*. "Agents for *The Moselm World*," *Moslem World*, August 1893, p. 8.

was at No 6 London St.[217] Went there between 4 and 5 and found that Col. O. had gone to Darjeeling at 4 but would return in two or three days. Ismail, another Moslem and myself drove to Eden Garden, where, after a walk about the handsome grounds, we said our prayers on one of the lawns, under a banyan tree. We then went and listened to the band for half an hour and encountered Abdool Latif who was quite talkative and pleasant. He was accosted, while I talked with him, by a Hindu judge of one of the courts who was bareheaded and clothed in the sheet of his caste. Latif gave him some advice—go in society more—get into the Viceroy's levee in Dec. Latif is almost an ass.

October 18, 1892

Between 8 and 9 Hajee Abdulla and a wealthy merchant and myself started out in a gharry to return the calls of Abdool Latif and his asinine son and two other prominent Mohammedans.[218] Out first call was at Latif's—the interior of his house was fully in keeping with his character. The man is consumed by an idea of his own importance and loves to pose as a leader and a man of great talents. His office, or the large room evidently used as a sort of office and library was a sight. Books and papers were scattered about in vast profusion for the purpose of giving visitors the idea that their owner was so deeply immersed in literature and law that he hadn't time to have his office arranged in an orderly manner. The fellow probably has more leisure time than he knows what to do with of all the self-sufficient, conceited people I have ever met Abdool Latif stands very prominent. And he's a schemer too—in his desire to be at the head of the Mo-

---

[217] Anagarika H. Dharmapala, 1864-1933, Sri Lankan Buddhist missionary who joined the Theosophical Society in 1884 and worked closely with the leaders of the organization. Silvia Cranston, *HPB: The Extraordinary Life of Helena Blavatsky*, 214-15. For a discussion of Col. Henry Olcutt, see below, December 10, 1892.

[218] Abdul Latif had two sons, Abul Fazl Muhammad Abdur Rahman, barrister-at-law of the High Court of Calcutta, and Abul Khair Muhammad Abdus-Subhan, Khan Bahadur. Roper Lethbridge, *The Golden Book of India*, 5.

hammedan community he has resorted to means of the most treacherous and despicable character and has succeeded in splitting up the whole community. We next called on his absurd son. He's funny, a sort of Anglicized Indian monkey who is harmless but very amusing and in his efforts to appear smart exposes his father's schemes. I trust that I will not be bored with further visits from this precious pair. About noon the merchant and I, accompanied by young John Mohamed went to Garden Reach and made a superficial inspection of the palaces of the late King of Oudh. Oh, what striking evidences of vanished magnificence and English rapacity and greed. Having driven the king from his throne and sequestered him and his family and dependants at Garden Reach, as soon as he was dead the government gobbled up all his property, leaving his family with small pensions and stripped the palaces of everything worth carrying off even to the marble tiles from the floors and fountains, the fine zoological collections and various statues and ornaments. The whole place presents an appearance of desolation depressing in the extreme. About 3 we crossed the river to the Botanical Garden, a most attractive place containing the famous banyan tree. The garden is most attractively laid out and the level well-kept drives flanked by great rare trees are most inviting. Beautiful vistas of roadway, lawn and rich green foliage. Pretty, but badly kept lakes, and well-kept conservatories. The banyan tree is a wonder. Parent trunk 42 feet circumference and area of top over 300 feet across. Hundreds of offshoots. Pleasant shade. Trunks entwined by orchids and vines. Returned about 5 to house of Hajee Abdulla Hajee Abdul Wahed formerly one of the palaces of the King of Oudh where we had some most excellent ice-cream, cake and tea and after prayers we drove to Calcutta. In the evening, after dinner I went out for a walk having to take that idiotic boy who is in the stamp business, as a guide. He is a confirmed ass. Insists that he speaks English and is always ready to answer "Yes" to everything he doesn't understand. He thinks that is a very clever way to delude people into the belief that he comprehends everything. Samples—"What is that?" "Yes." No, I mean what is that, there?" "Oh, that is—" "Do the Mohammedans all close their stores at dark?"

"Yes." Presently we came to a number of Mohammedan stores open. "I thought you said that all Mohammedan stores closed at dark?" Calmly "Yes." "How far is it to the next street?" "Yes." "I don't like this climate at all." "Yes." Presently seeing that he didn't understand half I said I asked: "Is that woman a man?" "Yes" "But women are not always men are they?" "Yes." "Did you ever see a white black bird?" "Bird, yes plenty here." "What kind?" "Yes." That's the sort of fellow I had to dig information out of. I succeeded in buying a native cap and pipe, however, and returned home discouraged.

October 19, 1892

Started a letter to Judge Knott, intending to fill in the time until 7 a.m. when the editor of the Mohammedan Observer was to come and copy my lecture of last Sunday night.[219] He didn't come until 8 and then he was accompanied by a committee who invited me to lecture in the Town Hall. They were very much disappointed when they found that I had decided not to lecture any more in Calcutta. Urged and insisted but I held out. Irish brogue'. It being too late to go on with copying the lecture I started with Hajee Abdulla Arab and Hajje Abdool Wahed to call upon some prominent Mussulmans. Our first call was on Dewan Syud Ameer Hossein, who smoked his hookah and showed by his manner that he did not like to hear anything said against England. Had never visited Europe and didn't know anything of America. Called then on Mohamed Youssuf Khan Bahadur, a very affable, intelligent and earnest man.[220] He talked freely of his new work on Mohammedan law and consented to my publication

---

[219] The editor was probably Moulvi Mahomed Bodi-ul-Alm who was editor of the *Mohammedan Observer* in 1893. Abdul Latif, *Autobiography and Other Writings of Nawab AbdulLatif Khan Bahadur*, 289.

[220] Mohamed Youssuf Khan Bahadur served as a member of the Bar of the Calcutta High Court and was elected to the Council Board of Bengal in 1895. "News Notes," *Moslem World and Voice of Islam*, April, 1895, p. 2; Lethbridge, *Golden Book of India*, 351; and *Who's Who in India*, 8:104.

of it. Invited us to dine with him on Friday. Then called on Moulvi Abdul Jubbar. He talked very little but read us his entire lecture on Mohammedan morality. It was very well written and very interesting but as I had an engagement at 10 with Jan Mohammed Gangjee to see the Public Museum and as it was 10 when he commenced reading the performance made me a little uneasy. Fortunately the lecture was a short one and he got through with it in about 15 minutes when we made our escape. When we reached home I found Jan Mohamed and the editor waiting and I made an engagement to meet the latter at 2. Went with Jan to the Public Museum and was surprised at its extent and completeness. It is a large gray granite building covering a block with a fine court inside. All the departments are well-kept and arranged and have, what seem to be, complete collections. The collection of old Indian religious sculpture is exceedingly full and would be enjoyed by the student of the oriental religions. Buddhas, and Hindu gods innumerable, and inscriptions carved in stone. Splendid collection of minerals, butterflies, fish and deep sea fungi—corrals, shells, birds mammals, and extensive collection of Indian embroidery, carving, sculpture, iron and brass work, heads of the various Indian races,—Rajputs &c rich gold and silver anklets, bracelets, earrings, nose rings &c. Fine ivory carving for sale cheap. Monkeys, stuffed and neatly arranged to show the development of the dog face to a strong resemblance to the human face. About 3 Capt. Hugh Fraser called and brought me a letter from Ella and dear little Mamie and Lora.[221] It made me rather homesick but I was glad to get it. Capt. Frazer reported Bro. Serajuddin Ahmed very sick at Midnapore and invited me, on behalf of the citizens of Gaya to pay a visit to that place.[222] Had to decline. He seems to be a good man but

---

[221] Captain Hugh Fraser was a Muslim convert that was charged with involvement in a cheating scheme at about the same time Webb was in India. "Local," *Times of India*, November 25, 1892, p. 7.

[222] Moulvi Serajuddin Ahmed was a Eurasian convert who along with Hajee Abdullah Arab, had visited Webb in Manila in March 1892 concerning the American Islamic mission. Webb, "Preaching Islamism," 469; and Ahmad, *Mujaddid Azam*, 159.

the Moslems here do not like him and tell rather hard stories about him. But I find considerable feeling against the European converts who happen to be here. English sparrows on the asotea. In the evening after dinner took our host's horse and carriage, and went out to see the Hindu illumination in honor of their god Kali the bloody one who is supposed to cut off the heads of the wicked to appease some other God. The Hindu shops and houses were generally illuminated— the sweets shops on one or two streets presented very pretty appearance. Fire-crackers, rockets, red lights &c. Streets crowded with people of all religions. Small shows. Escorts,—the simple-minded stamp dealer, and the editor of the Mohammedan Observer, persisted in pointing out to me things which hadn't the least interest for me, such as colleges and public buildings and ignored utterly those things in which I was deeply interested. Have asked several times to see a snake charmer or fakir—have been told indifferently that there are many here but none have shown up yet. These people seem to think that what is ordinary and commonplace to them must be so to everyone else.[223]

October 20, 1892

Put in the greater part of the morning in writing—Abdul Luteef called about 9.30 and staid for over an hour talking in his self-conceited way. He wearies me very much. After his departure Hajee Abdulla decided to go to Bombay next day—very much disgusted at the turn of affairs. Abdul Luteef seems to be his nightmare. After tiffin wrote letter to Ella and intended to go out for a walk afterward but rain came up and didn't get out until after 4 when went in Hajee Cassim's carriage with Ismail to Mohammedan cemetery— or seminary, as Ismail called it.[224] Very pretty mausoleum

---

[223] Entries in the first journal end here and written on the last page are the five Muslim daily prayers and presumably their respective number of rakaats, or cycles of prayer: Fujr—2-2, Zohar—2-2, Asr—2, Magrb—3, Isha—2-3.

[224] Most likely Hajee Yoosuf Cassim, but could be Hajee Kassim Arif an associate of Nawab Abdul Latif. Abdul Latif, *Autobiography and*

with three Moulvis buried under it. Well-kept graves and pretty mosque. Hindus along streets preparing for their Kali feast day. At night had to wait in for that pestiferous Luteef who had engaged to bring a Hindu friend to inspect me. He proved to be Dr. Sambha C. Mookerjee, editor of "Reis & Rayyet," a very pleasant, well-informed and intelligent old man.[225] I could hear the fire-crackers and various noises made by the Hindus in their celebration out on the street and wanted to see the show but found myself a martyr to the cause. Must be seen and inspected by every cranky nuissance that comes along. Got to bed about 11 and the mosquitos kept me awake for an hour.

October 21, 1892

This is the Hindu New Year—cows frescoed,—house fronts decorated with colored paper and paints—fire-crackers, bowls and colored lights. The goddess Kali carried through the streets with tom-toms & pipes. About 7 Ismail and Abdulla Wahed took me out to the Kali-ghat, about 6 miles from Calcutta. Here is quite a populous town populated by Hindus almost exclusively. We went to the temple in which Kali is kept and found that she could not be seen until 8:30 so we started for the burning ghat. We saw, however, where the sacrifice is offered to Kali, the bloody pavement and smeared parts giving evidence of the cruel act. In one of the temples was a vacant altar, or what seemed to be it was too dark inside for us to see clearly. We were told that Krishna was kept there and that he was asleep but would awaken in about an hour. When we returned about an hour later there was a gaily dressed figure on the altar playing the flute with his right hand while his left was around a gaudily attired female figure said to be his wife. He had one leg raised as if dancing and we were told that he and his wife were amusing

---

Other Writings of Nawab AbdulLatif Khan Bahadur, 221. Ismail is unidentified.

[225] Dr. Sambhu Chandra Mukerji, 1839-1894, editor and author of several periodicals and books. *Reis and Rayyet* began publication in 1882. C.E. Buckland, *Dictionary of Indian Biography*, 305.

themselves. At the burning ghat we saw the trenches over which dead bodies are burned, but there were no dead bodies. In the little temple or arched shed were three healthy looking yogiis, two chatting merrily and smoking ganja, or Indian hemp, which has an intoxicating effect. We looked out of the enclosure, or, rather, went out through the opening in the wall and saw the canal through which the sacred river Ganges has been turned in order to give the pious a chance to wash away their sins every day. They couldn't go to the Ganges so they brought the river here, or a portion of it. When one is given up by the doctors they take him to the river, stand him in the water up to his waist and pour water into his mouth until he dies. He is then burned—water-soaked. Under the guidance of a half-naked citizen and a small boy we went to see the god Juggernath, or several copies of him in paint and clay. Small temple at bathing ghat—at head of steps leading to the river—pictures of several of the gods in picturesque attitudes—riding elephants &c ½ doz yogis, covered with dust, hair matted and twisted into rope like curls sat in porch of temple smoking ganja and chatting. Asked if they had psychic powers or siddhis said didn't know what that meant. Could do no strange things—were only dirty and good and gave up their whole lives to prayer,-—when they were not chatting or smoking ganja. We asked to be taken to an advanced yogi who could do wonders, and were taken down the bank of the river about 50 feet further where we found tall, black, naked yogi with beard twisted into ropes and hair matted. Whole body, face, head to foot covered with dust. Squatted on a mat feeding fire. He wouldn't talk. One bystander who spoke English said he could do wonders and had great powers but wouldn't exercise them because that was God's work. He looked to me like a hopeless lunatic. He too smoked ganja, and was not averse to accepting coppers. Urged wonders, but bystanders said that all the yogis who did those were in Benares or in the jungles—never in city—hid themselves. Walked through native bazar and saw several figures of juggernath and other gods. Near temple boy with tray of flowers horrified when I stooped to pick one up,—groaned and warned me not to touch them. Every time we came near after that he picked up his tray and held it away so that we could not touch it with our contaminating clothes or hands. Time precious as we had a breakfast engagement and said we could not wait until temple opened. Guides anxious to have us wait, assuring us that temple would open at 8:30 sure. I doubted it and we drove off—short distance guides shouted temple open and we returned. Not allowed to get near doors, doors opened scrambling crowd paid 1 pice each to go in and offer fruits and flowers to the terrible goddess. Caught only a glimpse of the figure as the room or cell was very dark. Started to go followed by crowd clamoring for coppers. Voracious heathens. At 11 went to breakfast at house of Mohamad Yussuf Khan Bahadur,—very intelligent Native lawyer—large library—writing book on Mohammedan law—Breakfast good—but rather awkward with forks &c. Mosque—then rest of an hour or so and off with Ismail and Abdulla Wahed to palace of Kumar Debandro Nath Mullick, wealthy Hindu whose grandfather was a rajah.[226] Has his own soldiers in funny old uniforms guarding door with guns bayonets. Palace all of marble with immense court. 90 kinds of marble in floors of dining rooms. Wealth of statuary and paintings all imported from Europe—all in disorder and dust,—chaos of magnificence—furniture...moth-eaten—dust everywhere— beautiful court with fountains,— iron sofas &c, all white. Marble flows everywhere. Hindu altar at one end of court. Repairing and cleaning. fine grounds with fountains—lake— Indian cranes—ostrich and aviary of marble. Luxury gone to decay being ruined. Then to private Hindu temple of Rai Budree Dass Bahadur "Mookim" or jeweler "to his Excellency the Viceroy".[227] A most beautiful gem. Marble temple with pillars inlaid with squares and diamonds or colored glass and mirror in fanciful designs. Marble floor, polished—Krishna in cell. Rich <u>very</u>. Krishna and wife on ele-

---

[226] Kumar Debendra Mullick, 1835-1894, head of the Mullick family from 1887 until his death. He was active in many of the public and private institutions of Calcutta. *Cyclopedia of India*, 3:330.

[227] Rai Budree Dass Bahadur, well respected Jain jeweler known for constructing the Jain Temple in Calcutta. Satish Kumar Jain, *Progressive Jains of India,* 26.

phant each side of steps. Large buildings for temporary accommodation of worshippers. Place for Nautch dance. Fountains, lake full of fish, flowers, plants—private praying room for Dass over gate—velvet cushions and rugs—elegant. Statuary and gas jets. Then to Moslem cemetery mosque for prayer and to Calcutta burning ghat. 3 bodies burning, one wrapped in sheet waiting turn. 1 nearly consumed, 1 half burned, 3rd fire just lighted. Two more bodies brought in wrapped in cloth. Bodies stripped and legs doubled up and tied, not to prevent party from kicking but to economize wood. Ghastly sight—smoke and smell—tasted Hindu for two hours. Attendants slow and talky, laughed and joked. Strictly business—no sentiment. One corpse that of old woman—son having his head shaved. Went home and changed clothes for dinner given in my honor at Indian Club by Hadjee Ismail Hassen.[228] Principal guest the Prince Jehan Kader Mirza Bahadur, son-in-law of the Prince of Oudh sat at my left.[229] At my right was the Nowab Meer Mahomed Ali, Both spoke English well.[230] Fine dinner—no formality nor dress suits. Very enjoyable.

---

[228] The Indian Club was founded around 1882 as an instrument of socialization between Indians and the British. Wilfrid S. Blunt, *India Under Ripon*, 115. Hadjee Ismail Hassen is not identified.

[229] Jehan Kader Mirza Bahadur, member of the Central National Mahommedan Association and an associate of Nawab Abdul Latif. Abbasi, *Annals of the Central National Mahommedan Association*, 19; and Abdul Latif, *Autobiography and Other Writings of Nawab Abdul Latif Khan Bahadur*, 219.

[230] Nowab Meer Mahomed Ali, ?-1894, the first Muslim elected vice-president of the British Indian Association. Also a founding member and vice-president of the Central National Mohammedan Association in Calcutta. Webb eulogized him in the March 1895 issue of *The Moslem World and Voice of Islam*: "In the death, last October, of Nawab Meer Mahomed Ally, the progressive section of the Mussulman community of Calcutta, India, lost one of its brightest ornaments. Such faithful and intelligent Mussulmans are a direct benefit to any community." "News Notes," *Moslem World and Voice of Islam*, March, 1895, p. 2; Maitra, *Muslim Politics in Bengal*, 73; and K.K. Aziz, ed., *Public Life in Muslim India, 1850-1947*, 97.

October 22, 1892

At 10 a.m. went to the house of Nowab Meer Mahomed Ali for breakfast and had a quiet dull time—met several new faces—not particularly impressed by any of them. Returned about 1 and found juggler or magician, wife and child waiting for us—gave ordinary trick show—one or two very good—pulling thread out of throat—suspending woman by tongue—cutting off bird's head—very clever slight-of-hand, but nothing new. The mango tree trick very transparent—Cubberdar Monkey only a trick. At the conclusion of their performance in came two snake charmers with 1 very large rock snake, ½ doz large and small cobras some small, harmless snakes and three large black scorpions. One man allowed one of the cobras to bite him and then cured the wound with a peculiar root and a small black stone which adhered to the wound—blood came. Both teased the cobras and they struck at them several times. Rather a ghastly performance. In the evening a delegation of Shiahs called and presented their respects and a party of the hard-headed kind headed by Abdul Luteef's boy. They questioned me closely and Jubbar and one of the Moulvis displayed excessive fanaticism, narrow-mindedness and stupidity. It is the first case of unreasoning fanaticism I have yet seen.

October 23, 1892

Went to mosque as usual and after writing for an hour took a nap. Awoke to find that two young men from the Madrassah had called to pay their respects. Glad I was asleep, but hadn't time to rejoice much before two other callers put in an appearance, one of them exasperatingly stupid. About 10 went to a breakfast in the same block, in honor of the wedding of someone, but saw neither bride nor groom. Native music outside windows and oriental scene inside. All squatted on floor and nearly all smoked hookah. When called to breakfast in next room it was served on floor and eaten with fingers as is the custom. At one went to office of a merchant where we had chat and refreshments. Met there an Indian Mohammedan who had spent five years in the States trading.

Said he was well pleased with the country. I can't imagine why he returned. During the evening received cable from several brethren who desired to say good-bye. At 9:30 (Calcutta time) or (8:57 Madras) we started for Patna in a 2nd class car.

October 24, 1892

Arrived at Bankipore station, 6 miles from Patna at 7:15 (Madras Time) and were received by Moulvi Khoda Bukhsh Khan Bahadur, Gov't Pleader, who was accompanied by, several of his friends.[231] I rode to his house in a high cart with one of his friends. Fine large mansion in which was the Bukhsh Oriental Library of 3,000 volumes in Arabic and Sanscrit and quite large English collection.[232] Rare old manuscripts. After breakfast read for about an hour in Bosworth Smith's "Mohammed and Mohammedanism" and then accompanied by Bro. Abdulla and a Bankiporists started for Patna, six miles distant.[233] Patna, which comprises Bankipore and several other small towns is about 6 miles long and not more than half a mile wide and lies between the Ganges and one of its branches called the Julla. The northern part is built on the site of the ancient Hindu city of Pattaliputra which in its glory about the same time of the invasion of Alexander the Great. Stories are told of its grandeur and remains of its palaces are sometimes found. Now the house of poverty and wretchedness. Dirt, squalor and misery everywhere. From Bankipore to Patna one long street—native shops,—houses frescoed with dung-cakes. The Ekka, the oddest vehicle I have seen yet,—made of bamboo frame and canopy on two wheels with decorated horn or cone 2 feet

---

[231] Moulvi Khuda Bakhsh Khan Bahadur, 1842-1908, presided as chief justice of the Hyderabad State High Court (1895-1898). Naresh Kumar Jain, *Muslims in India: A Biographical Dictionary*, 2:13.

[232] During the period of Webb's visit the collection of the Oriental Public Library totaled closer to 4,000 volumes. Ibid.

[233] R. Bosworth Smith, 1839-1908. *Mohammed and Mohammedanism* was published in 1889 and for the time period was considered rather sympathetic toward Islam.

high on horse's shoulders. Harness "arrayed like one of these."[234] Called on the owner of "Patna Institute Gazette" but he was out—nephew, an embryo barrister was in and produced the manager of the "Patna Institute Gazette" a very obsequious little man who greeted me most cordially. After short chat we went to call on one of Bro Abdulla's old friends whom he had met at Mecca. He was an old man, apparently about 70 and almost toothless, but very wealthy. He took us to his private siesta room at the top of one of his buildings which overlooked the Ganges. Here a most beautiful vista presented itself. The room was about 100 feet above the river and was surrounded by a balcony from which a magnificent view of the country could be had. This was my first sight of the Ganges so sacred to the Hindu. It was apparently about a mile wide at this point and very shallow and the opposite bank spread away for miles on either hand a level fertile plain bordered in the dim distance by a dark fringe of palm trees. In the middle of the river opposite to us was a long, irregularly shaped sand-bar on one portion of which about a mile distant was a party of Hindus burning a dead body. This spot, perhaps, was chosen because of its being near the centre of the sacred river. On our side were high walls running to the water's edge and forming parts of residences with balconies where the cool breezes can be enjoyed evenings. Our host made some tea for us heating the water in an urn bought at Mecca and serving it in cups and with sugar bought at the same place. This was done in honor of me. The tea was badly made, but it pleased the old man to have us drink two cups of it. We started for home about 2:40, the urn being awfully hot. Mother lying by the roadside in dirt nursing her baby. Number of women in rags and very dirty. Saw new mosque being built by wealthy Mohammedan lady. In the evening had large number of callers, one an F.T.S. Found

---

[234] A saying borrowed from the Bible. "Consider the lilies how they grow: they toil not, they spin not; and yet I say unto you, that Solomon in all his glory was not arrayed like one of these." St. Luke 12:27 (King James Version): or "And yet I say unto you, That even Solomon in all his glory was not arrayed like one of these." St. Matthew 6:29 (KJV)

that I must sleep on bed made on old frame sunk about 10 inches from floor on my own shawl, and no mosquito bar. Very poor entertainment. Slept 3 hours, mosquitos nearly devoured me. Mistake our coming here due to stupidity of Abool Hassan in Calcutta urging Abdulla after I had told him we would not come.[235] Abdulla was taken in and I had to suffer. The people here did not want me.

October 25, 1892

Left for Benares at 7:30 A.M. and arrived at 12:15—on the way beautiful birds a kind resembling paraqueets, or loros, perched on telegraph wires. Women and men at work irrigating fields in peculiar ways. Mud villages. Changed car at Moghal Sarai. Crossed Ganges soon after leaving Saronapore where 2 soldiers got in car.[236] River very low broad field of sand (yellow) miles in extent. River gradually filling up. Crossed 2nd time opposite Benares. Here more channel and deeper but great banks of sand on side opposite to city. Viewed from iron bridge city presented imposing appearance but entering it, dilapidated buildings and mud houses give it a wretched appearance. Expected to see sombre but clean and attractive city—however, same as all the other mud house cities. To Dak-Bungalow—Old temple Bishmat[237]— King of Delhi came here 250 years ago. Abdul Gaffur guide—Sample "How did get it? 250—King of Delhi Alim Geer"—"Golding Temple"—Jung Bahadur temple Nepaul. "Golden Temple" of Ranjeet Singh of Ramjat built it (200 years old. Adabri) 250 yrs ago Ranjeet Singh gave plates. Ignorant guide—cow temple—dirt—crowd clamoring for bucksheesh. Vagrants of the most degraded type. Guide says they worship their prick. Lingham worship. "Benares, very

---

[235] Abool Hassan was the brother of Moulvi Khuda Bakhsh and a native of Patna, he joined the Calcutta High Court in 1887 and held several positions within the court system. His philanthropic ventures included the founding of the Calcutta Mahomedan Orphanage in 1895. *Cyclopedia of India*, 2:164-65.

[236] Possibly Serampur.

[237] Should be "Bishnath."

famous river side if see yogi very famous the see mosque famous river side. Squalor and wretchedness everywhere. The most persistent beggars in the world. Children being trained for beggars,—boy slapping his stomach—group followed us to river where we took "Bujjeser". Temples full of mud—temples & long flights of steps. Rajah of Jeypore—built "observatory" 425 years ago. One King of Benares built ghat, Mussulman caste—"He dead got." 1,000 yrs ago Nepaulese temple—fine carving—smutty figures—sexual worship—had been magnificent—must have been carved hundreds of years ago. Ruins of magnificent works of art. Numerous temples on side-hill where in the ruins vagrants and religious fanatics swarm like rats. Groups lying on ramparts enjoying cool breeze and beautiful vista across the river. Burning ghat on river bank—burning bodies at water's edge. Temples wrecked by water washing away earth. Houses piled one above another on the hill-side. Mud god—"All poor mans—3 days they stop here and then finish." Ancient mosque built by Sultan of Delhi 251 years ago—pigeon dirt. Guide had to stop to smoke—politely asked us to sit down till he had finished. Went to look for yogi—of course guide didn't know where there was one, although at first he declared that he knew of lots of them. After he had enquired of Hindus one of them directed him to small temple where we found 6 men who were said to be Brahmins. They did not know of any advanced yogis but sent one of their number out to get one. He returned with a Hindu who spoke English well and was unusually intelligent. He had heard of Col. Olcutt and H.P.B. but had never seen them nor read any T.S. literature.[238] Didn't need to be said when he had much better in Sanscrit. He knew of an advanced yogi. Failure. Guide piloted us to tomb of Mohammedan Moulvi, then to mosque then home or to Dak Bungalow. Guide said "That church Catholic—that chapel church Roman Catholic, yes, Roman Catholic that church." The fellow is useless for anyone who desires information and can speak only English. Passed

---

[238] Helena Petrovna Blavatsky, see below, December 10, 1892, for a discussion of both her and Col. Henry Olcutt.

through narrow street—shops on 1ˢᵗ floor and Nautch girls up stairs. Faces peeping out of windows and over balconeys.

October 26, 1892

The guide put in an appearance at 5:30 and began to try to talk Bro. Abdulla into more expense. Small donkeys with big loads. Monkey temple—a dozen monkeys and a regt. of Bucksheesh vagrants—must take shoes off but goats and dogs wander inside at will—I had put on clean stockings that morning. Guide "Queen of Buhul that"[239] "Did she build the temple?" "Yes, Queen of Buhul, large temple." "Does she live there?" "Yes, Queen of Buhul, Queen of Buhul she her temple." "Is she in there now" "Yes, not famous temple— other more famous." "When did she die?" "Yes" "I say when did she die?" "Oh, hundred years ago." "Then she is in there now alive and well?" "Yes Queen of Buhul." Left Benares at 10:17 and had to change cars at Mogahl Sarai for Bombay— train 2 hours late—wearisome wait. At Ahraura Road flowers on side of station for 100 yards. Good stations—bad coaches—After leaving this station country began to grow hilly—at Chunar quite rolling and rocky—hill to right of station apparently fortified—castles, turrets. Hills and stunted growths to left. Linghams in fields. Birds resembling paraquets but not as large as wrens. Not monotonous. Herd of camels under trees near Meja Road (station).

---

[239] Uncertain reference, possibly the queen of Sultan Buhlul Lodi, ?-1489, founder of the Lodi Dynasty of the Delhi Sultanate. Majumdar, ed., *The History and Culture of the Indian People*, 6:139-42. In a variant description of his visit to Benares Webb referred to her as the "Queen of Bulbul," but that was still an uncertain reference. "Looking for a Yogi," *St. Louis Republic*, December 12, 1892, p. 3.

V.

BOMBAY

October 27, 1892

Daylight very cool and drizzling rain—country began to grow hilly with range of mountains on either side and land heavily wooded—passed through rock tunnel about 100 yards long. Nests of tailor birds—rain the greater part of the day. Broad river with clear water rushing over rocky bottom and flanked by thick foliage near Bir—a very pretty scene. Received telegram from Bro. Budruddin telling us to stop at Kalian[240]—telegram received about 7 at one of the way stations.

October 28, 1892

Awoke at 3:30 and at 3:50 took wash and had just finished my prayer when the train stopped at Kalian. Bro. Abdulla had not had time to wash—it was then 4:20. As the train stopped Bro. Budruddin Abdulla Kur appeared, accompanied

---

[240] Budruddin Abdulla Kur, 1865-?, elected member of the Municipal Corporation for Bhuleshwar Ward and Justice of the Peace in Bombay. He was also an auditor for Anjuman-i-Islam. In 1891 Webb began a correspondence with Kur that lead to Hajee Abdulla Arab's involvement with the Propaganda. Kur was also the secretary of the committee in Bombay established to raise funds for Webb's mission. *Representative Men of the Bombay Presidency*, 95-97; and Webb, *Islam in America*, 68-69.

by 4 other Mussulmans, and gave us a most cordial greeting. I had a refreshing bath and about 7 we had tea, eggs, bread and fruit in the station restaurant. Soon after we had finished the train came along and we started for Bombay. Got into a compartment marked "Reserved for ladies" and instead of changing the sign board, which was hung on a hook on the outside of the car, to the next compartment, which was occupied by two Parsee ladies and their escort, the Guard, with true British stupidity ordered us into the next compartment and put the ladies into ours. When we arrived at the station of Bombay I was given the most cordial reception I have yet had. Hundreds of Mussulmans pressed forward to shake hands with me while a long line of school boys sang a song of welcome in Urdu. Then their teachers and others hung garlands of flowers about my neck and filled my hands with beautiful boquets. The people crowded about me so that it was with the greatest difficulty that I was escorted to the carriage in waiting which conveyed me to a handsome residence which I was told was at my disposal during my stay in Bombay.[241] When the parlor was well filled and I had shaken hands with all present one of the brothers read an address of welcome to me to which I responded rather briefly as I was very tired after my long ride in those awful English cars. The company dispersed about 10:30 and I had a rest of an hour, and a light luncheon after which we drove to the mosque which is the largest I have yet seen. There was a great crowd present and at the conclusion of the prayer they pressed about me in a way that threatened to suffocate me. I could not shake hands with all of them and several kind brothers made a passage for me through the throng so that I was enabled to reach my carriage. The crowd was so great about the vehicle that it had to pass through them to get started. Reached house pretty well tired out. After tiffin had a short nap and went out with Budruddin in his carriage to Am. Consulate to get mail. Received a good package of letters, two from Ella, and one from Bessie. Returned home and found that Mo. Kadri Bey had called, with his vice Consul

---

[241] Webb stayed in a bungalow at Chinchpoogly. "The Propagation of Islam in America," *Times of India*, October 24, 1892, p. 5.

Husney Bey, and left his card.[242] Had dinner, read my letters, and went to bed. From the general view I have had of Bombay it seems to me a much better regulated city than Calcutta. I certainly should prefer Bombay as a place of residence. Picturesque houses in Hindu Quarter.

October 29, 1892

Wrote to Ella and Mr. Carr,[243]—went to P.O. and posted Ella's and to Marck's & Co's to buy a pair of eye-glasses,— was duly robbed, of course—eye-glasses which sell in the States for 75¢ and $1.00 they asked 12 R's for or $ about $5.50—generously consented to sell them to me for 10 R's or about $4.75—original cost about 40¢—Saw Hindu women at door of house (about 12 or 13) crying and clapping hands—grief on account of death of someone inside. Soon after saw about 20 Hindu men with sheets wrapped around them marching through street and howling for death of someone. Oriental groups on streets—active scenes—curious method of drawing up skirts of females. Two-wheeled passenger carts drawn by oxen. Went to Apollo Bunder and looked off on harbor—prayed Magrib on bank of bay, and drove to music stand on the Esplanade where we heard the band play.[244] In the evening had Kadri Bey the Turkish Consul and his Vice Consul Husney Bey to dinner.

October 30, 1892

Put in the greater part of the morning on my trunks—reporter of one of the papers—not a very bright young man,—called,

---

[242] Muhammad Kadri Bey Effendi, Turkish Consul-General. *Times of India Calendar Directory*, 726.

[243] Alfred Carr, Webb's longtime friend and associate. As superintendent of the Insurance Department of the State of Missouri he had written several letters of recommendation for Webb's appointment as consul. Letter from Alfred Carr to President Cleveland, 16 May 1887, "Letters of Application and Recommendation."

[244] The Apollo Bunder was part of the European quarter; it was the landing place for boats in Bombay. *The Imperial Gazetteer of India*, 8:399.

and I gave him a point or two which he will probably not use. About 4:30 went out with Budruddin Kur to look for Dady Burjor's father but failed to find him[245]—drove along the beach road—crowds of men women and children lounging or playing on the commons near the sea—said our Magrib prayers on the shore and drove to Apollo Bunder. Rather an uneventful day and little or no disposition shown to show me the points of interest. No one seems to care whether I see the city or not. Plenty of promises but not fulfilled.

October 31, 1892

Another very quiet, uneventful day—put in the most of it writing letters and worrying about that cast-off clothing matter. In the afternoon went to call on the Turkish Consul and had tea and prayers in his office—went to American Consulate and got two letters both Indian—none from home. Took the usual drive to the bunder looked at the harbor—it was all there yet—said prayers on the bunder—walked up and down once and then drove home, had about 30 Mohammedans to dinner and we all sat on the floor and ate with our fingers in Eastern style. Sat for a while in the moonlight out in front of the house, while Hajee Abdulla and Hassan Ali talked with the guests about the American mission.[246] Was in bed about 9.

---

[245] Possibly D.S. Dady-Burjor who was later Webb's agent in Hong Kong for his publication *The Moslem World*. "Agents for *The Moslem World*," *Moslem World*, August 1893, p. 8.

[246] Moulvi Hasan Ali, 1852-1896?, born in Bhagalpur, began his career as a teacher and then became headmaster of a local school in 1874. He wrote several books on Islam and in 1886 turned to Muslim missionary work and began lecturing throughout India. He is credited with the conversion of hundreds of Hindus and thousands of Muslims who returned to their religion after hearing him preach. In 1894 he met with Mirza Ghulam Ahmad and joined the Ahmadiyyat movement where he was appointed as a Khalifa. Ahmad, *Mujaddid Azam*, 158 and 351-59.

November 1, 1892

Arose at 4:30 Budruddin Kur promised to come yesterday morning to take me out but failed to do so and last night promised to come this morning before 10. Sure. Of course I knew he wouldn't come—they rarely do—so a little after 9. I went to the Post Office and mailed some letters and papers. Returned at 10:30 and found, as I expected, that Budruddin had not been here. These Indian people are a good deal like the Spaniards—prompt to promise but slow to fulfill. In the afternoon went to see a Moslem mystic Shaikh Abdul Rahim and had a very interesting visit. He has become sensitive, clairvoyant and intuitional; he predicted full success for the American mission and said that he saw large numbers of Mussulmans in North America—also saw Mohammed on my right hand and Omar on my left. His place was up four flights of dark stairs in a bad smelling smoky tenement. Rather repellant surroundings. A very good man, apparently. Had about 10 or 15 Mohammedans to dinner. Was invited to attend a Mowlood but dinner was so late that I could not go—man came after me but there was no gharry. Just as I was about to get into bed a gharry came but it was too late.

November 2, 1892

Went with Bro. Abdulla and Hassan Ali to call on a man named Soleman—friend of the deep villain who called yesterday to question me. Afterward went to Victoria Garden,—small collection of animals and empty cages—tiger and bear cages very good with rocks in them and grottoes—Came home and found Sayyad Fakhruddin Abubakar Edroos, Prof. of Persian in Elphinstone College—had pleasant chat and Dady-Burjor's father came—then a Kurd and Budruddin's father.[247] After Asr prayer went to the Royal Asiatic Library in the Town Hall building—75,000 volumes, well arranged. Went to bunder and then to prayer, after which went to the house of Hajee Yoosuf Cassim, a rich ship owner, where a

---

[247] Budruddin Abdulla Kur's father was Abdulla Allysaheb Kur. *Representaive Men of the Bombay Presidency*, 95.

reception and dinner were given in my honor by the Anju-man-Taide-Musulmanan-e-Jadid of Bombay.[248] About 250 persons present. An address of welcome was made and afterward presented to me printed on Satin and enclosed in a handsome box.[249] I responded for about half an hour and then we went to dinner. After dinner Bro Hassan Ali addressed me in behalf of the society and I responded. I was decorated with a beautiful garland of flowers and was given a large boquet over which was sprinkled rose water. Returned home leaving Bro. Abdulla to talk to some of the leading members.

November 3, 1892

Beautiful cool morning—almost like a late September morning in Missouri. Went to Post Office and posted a letter to Ella, then to American Consulate and had long chat with Consul Henry Ballantyne.[250] In the afternoon went out with Badruddin driving in our carriage to the usual place—no change—his main idea is self-aggrandizement—thinks it makes him popular to go out with me. Poor fellow, he is to be pitied and prayed for. A very quiet day and evening.

November 4, 1892

Early in the forenoon received a call from 4 Shias,—neat turbans—very glad to see me and spoke the most encouraging and congratulatory words. Mr. Edroos brought friend—intelligent young barrister who invited us to dine with him

---

[248] Hajee Cassim's bungalow was located on Grant Road. The Anjuman-Taide-Musulmanan-e-Jadid of Bombay was headed by Moulvi Ubaidullah, President and Ghulam Muhammad Munshi, Secretary. "Dinner and Address to Mr. A.R. Webb," *Times of India*, November 3, 1892, p. 6.

[249] For the text of the welcome address, see Appendix F.

[250] Should be "Ballantine." Henry Ballantine, 1846-1914, as well as being consul he authored several books, including: *On India's Frontier; or, Nepal, the Gurkhas' Mysterious Land* (1895) and *Midnight Marches Through Persia* (1879). W. Stewart Wallace, *A Dictionary of North American Authors Deceased Before 1950*, 24.

to-morrow evening. Young Parsee called with permission to admit us to Tower of Silence, obtained through influence of Dady-Burjor's father.

> "Hazrat Mohammed.Sallalaho Allehe Vasa-lem. Khoda Hafiz."[251]

> "Allah hower sallay wa sallem wazid wa bar-rick allah sayed denar Mohammed un zerrce zahte wassirisk sharee fee syenul asmar was-sifahte wa la allahee assahabayhe Wassul-lim."[252]

Received a number of callers—went to mosque at 11:30 and after prayers Bro. Hassan Ali made a very effective speech and I shook hands with the people. More callers in the after-noon. At 4:30 went out riding with Budruddin and a young man from Zanzibar whose father is a rich merchant. This young man is a greater liar than Budruddin,—a more artistic, picturesque liar. He told of a shark killed off Zanzibar that had 36 human skeletons in his stomach. Budruddin blushed and seemed to feel that it was hopeless to try to best that. Then the prevaricator rounded up by telling of a fish he saw in the Arabian Sea, 80 feet long. I told about a shark killed off Coney Island that had the skeletons of 47 horses and 35 full sets of harness in his belly. This wearied the Zanzibar man and he became interested in passing objects. When an Indian or Englishman tries to outlie an American he has a large job. Came home and found Bro. Hajee Abdulla and Hassan Ali waiting for us rather impatiently—they were hungry.

---

[251] "Hazrat Mohammed. Sallalaho Allehe Vasalem" translates from Ara-bic as "Muhammad, may the peace and blessings of Allah be upon him." Hazrat is an Urdu title of respect given to great men. "Khoda Hafiz" translates from Urdu as "God keep you!" John T. Platts, *A Dictionary of Urdu, Classical Hindi, and English*, 487.

[252] This is an Arabic prayer that appears to have been recited by someone with an Urdu accent and phonetically transcribed by Webb. The parts that could be ascertained translate to asking God to send blessings on the Prophet Muhammad, his family, and his companions.

November 5, 1892

Had an experience with Anglo-Indian newspaper management which was amusing as well as exasperating. Called at
the office of the "Times of India" to get 6 copies of the 3rd
inst—several half-castes who spoke English were present—
was told that I would have to come in again at 10 as the man
who had charge of the papers was not there yet. Asked if it
was not possible to buy the papers before 10 and was assured
that it was not. This is a morning paper. Went to the office of
the "Advocate of India" an evening paper and was informed
there also that I would have to come in again at 10 as the
gentleman who sold the papers was not down yet. Papers run
on the principles that govern the Anglo-Indian press would
not last long in America. Called a boy to take my coat to the
tailors to have some small repairs made and to press the rear
part. This was at 7 a.m.—Boy said tailor would not have iron
ready before 12—consented finally to go and try to have the
rips repaired. Two hours later asked coachman to go get me
some newspapers—hadn't had his breakfast yet—boy finally
consented to go. Received a call from the Shaik in the afternoon and after Asr prayers went to the Hajee Cassim Yoosuf
house to the "Mowlood" but remained only a few minutes.
Budruddin was anxious to get away from the censure of the
man whom he had promised that he would have me at the
"Mowlood." As usual he forgot his promise and I accepted
another invitation. We went to the seashore and listened to
the band for about half an hour and then went to the house of
Nawab Zada Nasrullah Khan of Sachim where we had been
invited to dine.[253] Found the Nawab to be a monkey dressed
up uncomfortably in European clothes and most disgustingly
Anglicized. The poor fellow suffered torments in his uncomfortable dress and the perspiration pored off his chubby face
in streams. A more consummate ass it has not been my misfortune to meet in the East. A young man of good family,
liberally educated and possessed of ample means, a fawning

---

[253] Should be "Sachin." Nawabzada Sidi Nasrullah Khan, ?-1924, was the
son of Nawab Sidi Muhammad Ibrahim Khan. Dr. Ken Robbins, Researcher, e-mail message to editor, September 2, 2004.

lickspittle to the English who despise his whole race and use him only for their own pleasure—a deserter from his ancestral religion and social customs and a pitiable object generally. A briefless barrister who takes pride in showing that he has adopted British weaknesses. His cousin the Nawab of Junjera dressed in a bright canary-colored silk coat and white pants and decorated with diamonds was presented to me.[254] He was a tall young man with a countenance that could not be accused of over-intelligence—one of those descendants of ancient Indian royalty who have little left of the old nobility and who are gradually dying out or being crowded out of their property by the English. Just before we left he hung around my neck a long gilded paper necklace a fit symbol of his personality and condition in life,—cheap tinsel. Two wealthy merchants and a Persian professor were also among the guests. The dinner was very good,—a very fair imitation of an English spread but badly served. The Indian cropped out in spite of everything. The host's attempt to imitate the English small-talk was oppressively tiresome. Why in the world can't these people have some manhood and independence, and cultivate a true manhood, instead of following the brutalizing habits and customs of their conquerors? After dinner, Bro. Hassan Ali made a telling speech in Urdu. On the way home we encountered a Mohammedan wedding procession which smacked strongly of the Ancient Indian customs. A long line of carriages flanked by torch bearers, and containing friends and relatives of the prospective bride and groom was interspersed by bands of natives beating tom-toms and blowing pipes, while groups of young men rattled a curious looking concern, apparently a stick and chain of rings of iron. Relative followed on foot and boys gaily dressed on richly caparisoned horses. The groom completely covered by flowers, so that his face could not be seen, rode a horse followed by men bearing long glass cases, filled with flowers. He was going to the house of the bride's father to meet her, presumably, for the first time. On the following day he would take her to his own house. Just ahead of him

---

[254] Should be "Janjira." Nawab Ahmad Khan Sidi Ibrahim, 1862-1922. *Who's Who in India*, 7:21.

were two grotesque dancers who twisted their bodies in a peculiar way to the accompaniment of tom-toms and cymbals. We retired at 11:30.

November 6, 1892

Arose at 5:30 and prepared to visit the Towers of Silence with Dady-Burjor's father and Munshi.[255] Of course breakfast was late and the coachman and groom were late. When I reached Dady-Burjor's place Khetwady Back Road 12th Lane it was 7 o'clock and he had been waiting for me since 6. Stopped and took up Munshi, at his house near the Portugese church. At the foot of the hill leading to the Towers, we stopped at a toddy shop and I tasted the native toddy or coconut sap, for the first time. The government exercises a supervision over the collection and sale of the stuff and derives a very substantial revenue from it. Earthen jars are attached to the cocoanut trees up near the fruit part, and the bark is punctured or scratched until the sap or juice exudes. This drips into the pot which hangs to the tree all night or until it is full. It is then removed and another pot put in its place. It is called toddy, as well as arrack. When fresh from the tree it is sweet palatable and harmless, but when it has fermented for some time it has a sourish, malty taste and is intoxicating although not so virulent in its effects as whiskey. It is more like beer in its effects. The towers of Silence are approached by a flights or series of flights of stone steps through beautiful groves and under a viaduct which leads to the aristocratic European residences on Malabar Hill. On, or near the, top of the hill is a Hindu Temple, much to the disgust of the Parsees. From the Parsee prayer temple nearest the entrance gates a most beautiful view of Bombay can be obtained. The city and surrounding country are spread out like a map from Malabar Hill and the rocky islands in the harbor to the fort and Calaba a point stretching far out into the Arabian Sea. Every tall mill chimney and turret was visible although a fog

---

[255] Munshi is a common Indian name. Possibly Munshi Shaik Ahmed or Munshi Bauker Abdul Gani Muhimtulay, both are listed as having attended Webb's lecture in Bombay. Webb, *Three Lectures*, 23.

hung over some parts of the city and the whole aspect was very grand. We were not permitted to approach the towers nearer than 30 feet—beautifully laid out grounds and paths. Why should they be called towers, when only circular walls, 15 or 20 feet high. Saw model of towers—5 towers— vultures perched on the walls, dreaming in the sunshine and waiting for a corpse. Fierce, disgusting birds. Pleasant ride back. While writing in my room Hindu with one-stringed fiddle and dancing girl about 6 years old—contortion dance and squeaky fiddle. In the evening 4:30 started for Parsee wedding at Albless building—was met at gate by Munshi— seated in row of Parsees—bridegroom with garland of flow- ers sat next to officiating priest—procession with band from bride's building to grooms, with three ladies, two bearing trays with clothes for groom even to shoes—ceremony at door of passing betel and cocoa nut, over clothes 3 times, breaking egg &c before admitted to building. Flowers and betel passed around to guests. I was given garland and cocoanut as mark of special courtesy. Groom went to door of groom's house and same ceremony as for clothing was per- formed. He was given new hat and a cocoanut. Procession of males was formed and passed to bride's house or the hall where ceremony was performed. Here we were presented with flowers, betel and another cocoanut on behalf of the bride. After groom had performed his ablutions, tied the sa- cred thread about his waist and said the prayer proper for the occasion, he appeared at the door of the ceremony room and asked permission to enter. Before it was granted the same ceremony performed at the other house was performed. En- tered and seated himself on chair or mat in centre of hall. Soon joined by bride who seated opposite—sheet be- tween,—right hands clasped under sheet—bound together seven times with twine. Sheet dropped each threw rice at the other and company clapped hands. Then seated side by side she at his left while two priests pelted them for half an hour with rice and recited prayers. Then brass tray brought in with milk which was tasted and ceremony concluded. Congratula- tions followed. The bride a pretty, modest girl of about 18. Dresses of the ladies present very picturesque and the whole ceremony very pretty and odd. On arrival home received letter from Ella and one from Bessie informing me of her marriage. Poor child, I'm afraid she has made a mess of it. Hassan Ali annoyed by bigots and fanatics.

November 7, 1892

Another uneventful day. Occupied myself during the forenoon writing to Bessie and Ella and reading my lecture "The Faith of Islam" to Hassan Ali. The result of the latter was an acrimonious argument which left me feeling very blue and ashamed the rest of the day. In the evening took dinner at the Islam club given in my honor. About 35 present. Well prepared spread excellently served. Club house very neatly furnished and admirably located on the sea shore. A very creditable feature of advanced Mohammedan life. It smacked rather too strongly of the English to be really pleasant.

November 8, 1892

At 8:30 a.m. Hassan Ali left for Poona and Hyderabad to prepare those places for our visit. I passed the greater portion of the day in writing my lecture for Thursday night. About 5 Budruddin took his usual ride to the fort and allowed me to accompany him.

November 9, 1892

Arose at 4 and after prayers and breakfast wrote until 9:30 and then started for the Elephanta Caves. Budruddin had agreed to start at 9 but with his usual stupidity left the house at 9:30—arrived at the bunder and found that the owner of the yacht which was to take us to the caves, and some others had been waiting there since 9. Of course B. had a lie ready and thus squared the matter with our host.[256] The yacht was a sailer and we had to depend on the wind. Between 9 and 10 a favorable wind was blowing but at 10 it began to die down. Had B. been on time we could have gone to the caves in less than an hour—as it was it took us two. The yacht was very

---

[256] B. is Budruddin.

comfortable and handsomely fitted up. Arrived at the Elephanta Island I was carried up a long flight of stone steps to the house where the man in charge lives—on the shoulders of four Hindus. One short man and 3 tall ones made my chair tip very uncomfortably. The sun was hot and before we had gone half-way up the bearers were puffing at a great rate. I stopped them and told them to rest, after which we proceeded. Arrived at the top we were received very cordially by an old Englishman who had served in the Indian wars, and who was put in charge of the caves by the Gov't as a sort of reward for his services. Here, at an altitude of about 125 feet from the sea, he lives in a small bungalow with his aged wife and with no one about but two or three Sepoys, and some half-naked natives who infest the stairway and squeeze pice out of visitors. The old man and his wide invited us into his bungalow which was supposed to have been built by the Portugese about 100 years ago—a small stone and chunam structure with a thatched roof. They were very cordial and talkative and said they would enjoy themselves in their quiet home but were unable to get all they wanted to eat. There was no regular means of communication between island and Bombay, 8 miles distant and they had to depend upon a native who sailed across in his canoe at irregular intervals and brought them provisions. The caves are wonderful—hewn out of solid trap rock as hard as flint they are supposed to have required the services of thousands of men, women and children for over 30 years in their excavation and carving. Paid in grain. The old man showed us about and described the sculptures dropping his "h's" most freely. The gods Brahma, Vishnu and Siva with angels attendants and lesser gods carved in wonderfully artistic manner. Three linghams in crypts. A spring of pure crystal water. The beastly, ignorant, fanatical Christian Portugese destroyed many of the beauties of the place. Put a cannon in the entrance and fired into the caves thinking that by blowing down the massive pillars the whole thing would collapse. Thought idol worship was wicked and the ignorant brutes wanted to break it up. They blew down several of the grand pillars and badly shattered others which have been repaired by the British government. Not restored. Massive stone elephant which faced one of the caves taken away by the British and put in Victoria Garden Bombay where it lies broken in pieces. The caves as they were when the Portugese arrived must have been grand and beautiful beyond description. There are two other caves but they are in a wrecked and dilapidated condition. There are also several small caves in various parts of the hill where figures of the gods are carved out of the solid rock. The island must have been a great place of worship for the Hindus many hundreds of years ago. We took our meals on board the yacht,—they were well cooked and served by our house servants. Had a pleasant sail back—caught a glimpse of handsome shrine built in a secluded spot over the remains of a Mohammedan Saint. Anchored of the bunder about 6 had dinner and reached home about 8.

November 10, 1892

Spent most of the day in finishing up and rehearsing my lecture for the evening. Mr. Dady-Burjor, Munshi and another Parsee called early in the forenoon and chatted for an hour. In the evening lecture on "Islam" at the Framjee Cowasjee Hall before about 400 people.[257] As the affair was wholly managed by Budruddin it was, of course, botched. The hall should have been full of people and would have been if he had made it an open lecture free to all; but he advertized it as for Mohammedans only and very few more than those who could speak English came. I was freely congratulated upon my lecture and its delivery.

November 11, 1892

Another dull, uneventful day. In the forenoon wrote to Ella and Mother and then slept until 11:30; took breakfast and went to mosque. Budruddin promised to come at 11 and ac-

---

[257] Named after Framjee Cowasjee Banajee, 1767-1851, a Parsi agriculturalist, who served in the government in many capacities and was known for his philanthropic ventures. Buckland, *Dictionary of Indian Biography*, 25. For text of lecture see, Webb, *Lectures on Islam*, 19-29.

company me to mosque—failed to come and I went with Haroun. Saw him at mosque,—the usual handshaking was done and we came home. Spent the afternoon in rearranging my trunks. Budruddin promised to come at 4 and take his usual ride. I waited until a quarter to 5 and went off without him. Met him and Haroun in Bombay—of course he had a plausible excuse. He is a great liar and as slippery and unreliable as an eel. They got into carriage and we went to Times of India office, bought six papers, and then drove to the bunder; had prayers and returned home where we had ten Moslems to dine with us. Bill of fare meager—the $600 seems to be getting low.[258] Where has it all gone.

November 12, 1892

Bro. Hajee Abdulla was out all night at work on the mission matter. He is a very good man; there are few in this world who would work as he is working, without any hope of earthly reward. Went fishing at 10:30 with Hajee Ismail on the yacht.[259] He promised to call for me at 9 and, of course, came at 10; the usual custom in this country. Started from Apollo Bunder at 10:35 and had a very pleasant sail to the fishing ground, arriving there about 12. Threw out lines and I am ashamed to say that I joined in the sport—caught four large fish, like rock bass—got tired of it after a time and lay down to sleep on the cabin house. After tea or about 3 o'clock, Hajee Ismail and our other companion got into the rowboat with five of the crew and rowed off to look for better fishing. I preferred to remain on board and enjoy the breeze and the quiet. Lay down for a time, then threw out my line but had no bites—then sat and meditated. Finally at a signal from the row-boat the men raised anchor and we

---

[258] The origin of the $600 is unclear, however, in his contract with Hajee Abdulla Arab, Webb was to be provided $200 per month in salary. This being the third month of the contract the $600 may have been his salary monies. "To My Oriental Brothers," *Moslem World and Voice of Islam*, January, 1895, p. 3-4.

[259] Possibly Hajee Ismail Sobani who along with his sons were influential members of the Anjuman-i-Islam. *Cyclopedia of India*, 2:325

drifted for about an hour until we came to the small boat—took Ismail and the others on board and set sail for Bombay with a fair wind. Hindu wouldn't wash Hajee Ismail's cup—Ismail said: "Koran says just same like dogs." Very much disgusted. Budruddin, met us with the carriage at Apollo Bunder and after a short drive came home to dinner. Found several people waiting to see me, among them Munshi Shaik Husein V. Shaik Chand Joonarkar I, a public preacher. A man with a name like that ought to be able to do some good in the world. He had with him a fresh young man who wanted me to know that he had joined 12 sects of Christians, had investigated it thoroughly and found it a fraud, was a Theosophist and was generally brilliant. I have seen some very queer specimens since I came to India and he is one of them. Went in to dinner and left them sitting in the reception room. When I returned they had gone, much to my relief. Boy Mohammed arrived from Dacca.

November 13, 1892

Received a call early in the forenoon from Dady-Burjor, Munshi and another Parsee. At 9 left for visit to Yussuf Soliman at his handsome residence on the sea-shore.[260] We reached it by a most delightful drive. It is a large, roomy but scantily furnished mansion situated in a finely laid out flower garden and having a beautiful view of the Arabian Sea. The shore is low and rocky and when the tide is low it looks very attractive. The road runs close to the shore. One can sit on the verandah of the house and look far out to sea, facing Aden Arabia. Hajee Yussuf gave us tea and some very delicious cakes and fruits, and then insisted that we should remain and take tiffin with him at 1. We accepted the invitation putting in the time, in the meanwhile, sitting in easy chairs on the verandah, and chatting; that is Bro. Hajee Abdulla! Hajee Yussuf, Budruddin and Hajee Ismail talked in Hindustanee and I dozed and finally fell asleep. When Hajee

---

[260] Possibly Hajee Yusuf Mohammed Sulaiman, vice-president and trustee of the Bombay committee to fund Webb's mission. Webb, *Islam in America*, 69.

Yussuf found that I was sleeping he ordered a cot and pillows brought for my accommodation but I did not use them. Our tiffin was very good and at 2 we started for Mahim, a small town about six miles from Bombay where there is the shrine of Ali, a Mohammedan saint who died about 600 years ago.[261] The drive thither was not charming as it led us through dense groves of cocoanut trees nearly the whole distance. The sun was hot and this shade was very acceptable. Saw Hindoo huts and villages where the natives seemed to be sunk to the last stages of poverty. Principal occupation gathering toddy—for every 21 rupees, Government exacts 20,—native gets 1. Object is to make natives buy English whiskey which is admitted duty free. Gov't also supplies prostitutes to soldiers, and exacts an enormous tax on native salt. Squalid scenes but very oriental. At shrine, oldest mosque in Bombay Presidency. Tank of water covered with green slime to render ablutions in. Children in front porch of mosque reading Koran, teacher supervising them, and hordes of beggars outside. Hindu as well as Mohammedan. Shrine well-kept. Met Mussulman from Bombay who took us to his house and gave us coffee and cardamom seeds.[262] After we had said Asr prayers we started for home and had driven along about half a mile when Hindu boy came running after us shouting for us to stop, said someone wanted to see us. It proved to be a Mussulman living in Calcutta who had come here for health on the sea-shore in the cocoanut groves. He had seen me in Calcutta, and knowing Hajee Abdulla wanted us to stop and take tea with him. We went into his bungalow which is situated in a dense grove of cocoanuts and papayas, and right on the sea-shore—a most charming and healthy place. It was the property of his grandson. We had tea and chatted and then resumed our journey. Passed out of the cocoanut groves across a swampy stretch to a rocky hillside

---

[261] Ali Maha'imi, 15th century Sufi scholar.

[262] Cardamom: "A spice consisting of the seed-capsules of various species of *Amomum* and *Elettaria* (family Zingiberaceæ), natives of the East Indies and China; used in medicine as a stomachic, and also for flavouring sauces and curries." *Oxford English Dictionary*, 2nd ed., s.v. "Cardamom."

on which were numerous squalid Hindu huts. The drive was very pleasant and gave us a fine view of Bombay. Arrived home in time for Magrib prayers. After dinner found that plans for leaving Bombay on Tuesday had been upset. The usual thing. Procrastination and postponement after all my arrangements had been made. Such a wretched, shifting, vacillating way of doing business is most exasperating. I never know what to depend on. I shall give a sigh of relief when I get out of this country.

November 14, 1892

Went to Post Office to mail letters and papers and thence to American Consulate; had a general chat with the Consul. Went to offices of "Bombay Gazette" and "Times of India" bought papers and received galley proofs of lecture.[263] Spent the greater part of day in putting papers in wrappers and correcting proofs. Was to have gone to Tyabjee's house at 4:30 but Budruddin was nearly and hour late, as usual.[264] Started about 5:30 and had a delightful drive along the seashore and up through Malabar Hill where beautiful residences nestle among the trees and give their occupants a fine view of the Arabian Sea. Tyabjee's house was magnificently furnished in European style and showed striking evidences of wealth and refinement. Above the entrance to the house was a stone asotea commanding a fine view of the Arabian Sea. We went up there just at sunset to say the Magrib prayer and the prospect was grand. The sun had just sank into the Arabian Sea

[263] Webb's speech was published in its entirety in the *Times of India*. "Islam," *Times of India*, November 14, 1892, p. 9.

[264] There were several prominent members of the Tyabji family in Bombay, but this appears to be Badruddin Tyabji, 1844-1906. British educated and adept at languages, he became the first Indian Barrister in Bombay. He held several positions within the Bombay Government and along with his brother, Camruddin, founded the Anjuman-i-Islam in Bombay (1876). His residence was called "Somerset House," belonging once to the former Commander of Bombay, Sir Henry Somerset, 1794-1862. It was located on Cumballa Hill near Malabar Hill. Nagendra K. Singh, *Encyclopaedia of Muslim Biography*, 5:377-79; and Husain B. Tyabji, *Badruddin Tyabji: A Biography*, 32.

leaving a soft effulgent, golden mist behind it and tinging the water to a subdued crimson which slowly faded into a grayish fleecy light and then into darkness. A gentle breeze stirred the surrounding foliage sufficiently to add life to the scene,—it was grand. After prayer we chatted for an hour and then went to dinner in the beautiful dinning room. The meal was deliciously cooked and perfectly served. After dinner we entered the magnificent drawing room where I was introduced to a small army of Tyabjee's nephews and cousins. On either side of the drawing room was a large room with double doors. Screens were placed in front or in the doorways, and in these rooms, which were dark, were assembled a number of Mohammedan ladies who had come to hear me speak. I addressed them upon the social systems of Europe and America as compared with that of Mohammedan countries but as I could not see my audience I was somewhat embarrassed and did not make a good speech. The host and the other Mohammedans, however, seemed to be very well pleased. We left at 9:30 and arrived home about 10.

November 15, 1892

About 6 a.m. went out for a walk and visited the Victoria Garden. Morning cool and pleasant, and the garden delightful. After breakfast put in the greater part of the forenoon in wrapping and addressing papers. In the afternoon received telegram inviting me to Poona. Went to office of Turkish Consul to see Hajee Abdulla and afterward telegraphed to Poona that I would leave Bombay at 10 a.m. Thursday. Had a drive, listened to band on the Esplanade and then came home to dinner. Another uneventful day.

November 16, 1892

In the morning took my 3 trunks to the American Consulate to store while on my trip to Hyderabad and other places. Saw in "Times of India" communication signed "A. Mohammedan" advising his brethren to keep their money for the poor of India rather than put it into the American mission— also a faint squeak from one of the missionaries declaring that my interpretation of St. John 8-58 was wrong.[265] Poor fellow he is like all the rest of the blind. Hajee Abdulla Mohammed and another man spent the greater part of the forenoon in constructing a letter to the Gazette to the effect that my visit to India was not for the purpose of collecting money. But it fell into the hands of Budruddin and did not get into the paper as it was calculated to deprive him of some of the glory of my discovery.[266] He has a perfect mania for notoriety and an utter lack of honor and fairness. Spent most of the day reading and loafing. In the evening made the usual trip to the Apollo Bunder,—home to dinner bed at 9.

November 17, 1892

About 7 received a call from Hosein Effendi of Mossul, Kurdistan, who claimed that 120 prophets were born in his town. Left for Poona at 10:36. Budruddin wanted to go to Poona with us but the expense was too much for him; of course he gave another reason which was a lie. Of all the gifted, prolific liars I have met in my life he is the high priest. Our ride to Poona was most delightful. Dady-Burjor's father got on the train at Byculla and rode with us to the next station. One of our Mohammedan friends accompanied us from the Victoria Station in Bombay to Goorla, about ten miles out.[267] At noon Haji Abdulla opened our lunch basket

---

[265] For the text of letter from A Mohamedan, see Appendix G. John 8:58: "Jesus said unto them, Verily, verily, I say unto you, Before Abraham was, I am."

Text of missionary's letter:

Sir,—May I be allowed to remind your readers that although the translation of St. John viii., 58, proposed by Mr. Webb last Thursday evening may be what he thinks it should be, still the Revised Version of the Greek Testament has the support of leading Greek scholars of the day, both in America and England.—Yours, &c. H. Lateward. "An American Convert to Islam," *Times of India*, November 16, 1892, p. 5.

[266] A letter concerning the matter was published in *The Times of India*. Webb, *Lectures on Islam*, 18. For the text of the letter and replies, see Appendix G.

[267] Possibly Kurla.

and found a nice roast chicken with potatoes and a number of fine fresh figs, and bananas, and we made a very nice lunch. At Karjat we took on another locomotive and began the ascent of the Bhore Ghauts or hills—they might very properly be called mountains. As we approached them, before reaching Karjat they presented a very curious appearance, having water lines as if they had been washed by a great sea which had slowly subsided leaving them to stand as monuments of a time of which no record has been left to man. We ascended at the rate of about 20 miles and hour over as fine a road-bed as I ever saw and before reaching Khandala at the top of the ghauts, passed through at least 25 tunnels cut through the solid trap rock. The view at times was simply magnificent as we looked down into the valleys and vast stretches of fertile country below. We were ascending constantly and at a height of at lest 1,000 feet looked down upon the paddy fields and villages covering an area of many miles before it merged into the hills in the far distance. The sun shone glaringly on the grand vista and with the yellow and brown of the dried vegetation made it unusually gorgeous. The rainy season was over and the cold season had commenced. This R.R. is a commendable a piece of engineering as any I ever saw and must have cost a vast sum of money. On our arrival at Poona we were received by a large crowd of Mussulmans—perhaps 500, and were taken in a handsome carriage to the bungalow assigned for our use. It was the residence of a wealthy gentleman who had gone away on a visit somewhere and we were left in complete possession. It was magnificently furnished and had a superb flower garden laid out in front of it while the porch surrounding it was decorated with rare plants and trailing vines. We could not have been more comfortably settled. After a brief interchange of courtesies between a number of the leading Mussulmans and myself and Bro. Hajee Abdulla, Hassan Ali and myself had been duly garlanded with flowers, as is the custom, and given tea and cake, we went out driving and said our Magrib prayers in a public garden. My first view of Poona impressed me very strongly with the beauty of the streets and the tasteful arrangement of the place. Bro. Hajee Haroun Hajee Jaffer accompanied us and after prayers took us to his store, or "shop," as he called

157

it.[268] He is a shining example of the self-made man. He began here 30 years ago as a book peddler, and subsequently opened a little shop for the sale of books. He has gradually increased his business until now his buildings cover a space of nearly a city block and he carries an immense stock of almost everything from a bottle of ink to a set of parlor furniture. He builds and sells carriages, has a tailor shop and rents whole outfits of household furniture to army officers. The Bombay division of the British army is quartered here and hence he finds this business profitable. He is very rich and is accumulating money rapidly. We drove home and after I had dressed, had dinner and a very fine dinner it was too. 20 guests had been invited but only 15 attended. Dinner a very quiet affair. Went to bed about 10. Was awakened at about 3:40 by hearing the words "Good gracious!" shouted in my ear.

November 18, 1892

After prayers, Hajee Abdulla, Hassan Ali and myself went out for a walk. The air was cold and bracing—the first really cold weather I have felt since I left the States 5 years ago. I got out a pair of woolen pants and a Prince Albert coat which I buttoned up tight. Felt like a new man,—the bracing air invigorated me and seemed to give me new life. I walked at a very brisk gait and did not perspire at all. We visited the falls or dam made for the Poona water works from a subscription largely given by Sir Jamsetgee Jeejeebhoy of Bombay and the botanical garden near.[269] The latter was most beautifully laid out with gravelled walks and well-kept flower beds, and terraces. Returned to the bungalow about 7 and had a very nice breakfast of milk, toast eggs and fruit. After breakfast Bro. Hajee Haroun Hajee Jaffer Yussuf called and took us

---

[268] Hajee Haroun Hajee Jaffer Yussuf was later Webb's agent in Poona for his publication *The Moslem World*. "Agents for *The Moslem World*," *Moslem World*, August 1893, p. 8.

[269] Sir Jamsetgee Jeejeebhoy, Baronet, 1783-1859, a Parsi who engaged in international trade and used his amassed fortune to engage in philanthropic work. Buckland, *Dictionary of Indian Biography*, 223.

for a drive through the town. Poona proper is a quaint old place and is the most oriental town I have seen yet. The little shops kept by Hindoos and Mussulmans were very novel and very dirty. We also visited the market and saw some very fine fruits and numerous piles of peanuts which grow here very plentifully and form one of the staple articles of food of the people. Very fine pomolos, and custard apples and papayas large and delicious. Poona is a great place for fresh figs. Poona was formerly a town of the Peshawars who hastily evacuated it on the approach of the British and have almost become extinct.[270] Some of their coin is still current and the building in which the small shops are situated are, many of them, servants of the Peshawar days. It is an odd quaint old town and full of interest. Outside of its limits is Poona, proper, the headquarters of the Bombay Army and the resort, during the summer months, of the better class of Europeans. We, or rather, Hajee Haroun, bought for us a lot of fruit which was taken to our bungalow. After visiting the town we went to Hajee Haroun's store and sat there for half an hour and then went to the bungalow. At 12 Hajee Abdulla, and Hassan Ali and myself went to the mosque for prayers. After prayers we went to the store of Hajee Soleman, and from there to Hajee Haroun's store to try on an overcoat I was having made there.[271] Hajee Haroun would take no pay for the coat but insisted that I should accept it as a present. I did so reluctantly. We returned the bungalow and stopped at the house of Khan Bahadur Kazi Shahabudin C.I.E. a very well-educated middle-aged gentleman and had a pleasant chat there.[272] He has dabbled some in Theosophy but is not very

---

[270]Should be, "Peshwa." Reference to the Anglo-Maratha wars (1775-82, 1803-05, 1817-18), which lead to the demise of the Peshwa (Maratha princes) in Poona. Mehra, *Dictionary of Modern Indian History*, 431-35.

[271] Possibly Hajee Sulaiman Ilyas, member of the committee to raise funds for Webb's mission. Webb, *Islam in America*, 69.

[272] Khan Bahadur Kazi Shahabuddin, 1832-1900, held several positions in the Bombay Government and served as Prime Minister of Baroda from 1883-86 before retiring. He was also vice-president for the committee raising funds or Webb's mission. Jain, *Muslims in India*,

clear on it. At 4 I was visited by a large number of Mussol-
mans and, was presented with an address printed on blue
satin and enclosed in a handsome sandal-wood box. Kazi
Shahabudin presided and made the presentation speech in
Urdu. I responded in English, speaking nearly an hour and
apparently pleasing the audience. My remarks were translated into urdu by Hassan Ali. A garland of roses and other
flowers was placed about my neck and I was given a sceptre
of roses and a fragrant boquet. We were sprinkled with rose
water and the company dispersed. After a general chat with
the few who remained we had dinner at 8 the Persian Consul-General and his vice-consul being among the guests.
There were about 21 persons at the table. After dinner I had a
chat with Kazi Shahabudin on Theosophy and spiritualism
and then went to bed about 10:30. The engine-driver who
came 300 miles to see me and wanted blood.

2:147; Lethbridge, *Golden Book of India*, 491; and Webb, *Islam in
America*, 69.

VI.

HYDERABAD

November 19, 1892

Arose at 5 and began preparations for our journey to Hyderabad. Took 8 o'clock train 2nd class A.M. and about ½ a dozen Mussulmans were at Station—among them Kazi Shahabudin and Hajee Haroun who had provided a nice luncheon for us and decorated us with garlands of flowers just before the train left. They regretted to have us leave. Bad arrangements at station for getting pilgrims on board—hundreds packed outside and admitted one by one through an iron gate while a boy punched their tickets. The usual blundering English customs prevalent everywhere. 4 Hindoos in our compartment but they didn't stay long. Had a disagreeable, dusty ride all day and at 10 p.m. arrived at Wadi, in the Nizam's territory where we changed cars for Hyderabad—these cars worse than the others.[273] No water—as usual. I traveled 1st class while Hajee Abdulla and Hassan Ali went 2nd as it was not considered the proper thing for me to go into aristocratic Hyderabad 2nd class. At Wadi found no light in

---

[273] The Nizam was Nawab Mir Mahbub 'Ali Khan, 1866-1911, ruled from 1869-1911, but under co-regents until he was of age in 1884. The Nizam was the most powerful native ruler in India and although he maintained autonomy over his territory he remained a friend of the British. Singh, *Encyclopaedia of Muslim Biography*, 3: 433; and Sheela Raj, *Mediaevalism to Modernism: Socio-economic and Cultural History of Hyderabad*, 18.

closet—Station-master's clerk surprised but thought I had better tell the station master about it. Saw native porter with lantern called his attention to the absence of lamp. Examined the car in open-mouthed wonder inside and out, but didn't know what he could do about it. Saw another man with lantern and explained the situation to him—he called a porter and said something to him in Hindustanee—latter examined car and seemed dumbfounded but was powerless to do anything. While he was considering the matter the station master—apparently half Irish and half cockney wholly soaked with whiskey—came along. I stated the case to him when he informed me that no lamps had been supplied. I didn't know what that meant so asked him. He did not deign to reply but said to the porter "That's all right, you go on." thus summarily disposing of the matter. Afterward ascertained that it was the policy of the Eng gov't to depreciate everything belonging to the Nizam's gov't with a view of making it appear that the Nizam's territory was badly managed and needed the intervention of the British gov't. Some day, when it has sufficiently demoralized the Nizam's following it will coolly pull him out of his position, send him away and quietly gobble up his property. That is what they did in Burma and wherever they have had a chance. Had a fairly good sleep and the compartment all to myself.

November 20, 1892

Was awakened at 4 by the shouting of the usual Hindoo at a station. There are always, at every station two or three mullet-headed Hindoos who seem to think it their duty to go along the platform yelling at each other and awakening every passenger on the train. They are employees of the railway in the discharge of their duty and are ordering one another about thinking that this impresses the by-standers with their importance. Human nature is a queer combination. Soon after daylight we came to a part of the country which presented a very strange aspect. It was covered with rocks of various sizes worn into boulders by the action of water or ice and seemed to indicate the extreme plausibility of the theory that this section was once covered by ice bergs which carried

162

these rocks along, wearing their edges against other rocks. In short it seemed a corroboration of what science believes to have been the conditions of the glacial period. The train was very slow and made a number of protracted stops while approaching Hyderabad. When we arrived to within about 100 yards of the platform quite a number of Mussulmans put in an appearance and some of them entered the car and greeted me cordially. At the platform their was a vast crowd in waiting to receive me and I was escorted to a handsome canopied carriage—evidently a state vehicle—and under an escort of mounted native soldiers was driven to the Nizam Club.[274] The cavalcade presented quite a showy appearance. A gaily caparisoned elephant was also in attendance and gave the scene a very oriental aspect. I was also struck with the general oriental appearance of the town as we passed through the streets. At the club I was presented to a large number of well-dressed, Europeanized Mussulmans, who were apparently of a higher order of intelligence and refinement than any I had yet seen. After tea and a bath I had an opportunity to write and also slept nearly two hours. In the afternoon about 4:30 I expected to go driving but a lot of thoughtless people came and deprived me of that pleasure as I had to remain at home and receive them. Had a general and somewhat unsatisfactory chat during the evening before dinner. There is apparently too much free-thought, skepticism and atheism here to admit of much good work. Club in finely located in a beautiful garden and as I have all the comforts I can ask for I shall probably be contented.

November 21, 1892

Weather chilly and pleasant. Wrote until 8 a.m. when the Supt. of Public Instruction came and took me to his house in Chuderghat to see his translation of the first three chapters of

---

[274] The Nizam Club was the first European style club in Hyderabad and was founded in 1883 by Syed Husain Bilgrami and Rafat Yar Yung. Saidul Haq Imadi, *Nawab Imad-ul-Mulk*, 124.

the Koran.[275] House furnished in European style with a profusion of pictures and a grand piano in the parlor. Photographs on table indicated that the Supt. had fallen deeply into English customs. While we were seated in his library, pretty young lady in jacket, and jaunty hat passed through greeting the Supt. pleasantly. No purdah here. Rather a queer fish is the Supt. After a chat of about an hour we went to the new Mohammedan library It is well arranged in a fine, commodious bungalow and has a good stock of books to begin with, largely Arabic. Came home about ten and worked on my lecture the rest of the day with occasional interruptions from callers. At 5 went out driving with Mehdi Ali, Hajee Abdulla and Hassan Ali.[276] After a short drive in the English quarter we went to the old walled city—the first really oriental city I have seen. The old walls, turreted buildings among the trees, narrow streets, little shops, arches over the street, elephants plodding along with bells on their sides and half a dozen people on their backs, gaily colored dresses and turbans, odd-looking bullock carts and gharries gave the place a novel and attractive appearance. Fronts of shops gilded and arched—dirt and vags everywhere—cheap show. Carriage passed containing prime minister arrayed in tinsel and span-

[275] Nawab Imad-ul Mulk, Syed Husain Bilgrami, 1842-1926, Director of Public Instructions (1887-1902) and appointed private secretary to Nizam in 1888. He wrote several books including: *Historical and Descriptive Sketch of His Highness the Nizam's Dominions* (1883-4) and *A Memoir of Sir Salar Jung, G.C.S.I.* (1883). Imadi, *Nawab Imad-ul-Mulk*, 125; and Singh, *Encyclopaedia of Muslim Biography*, 2:151-52. Location of his home should be "Chaderghat."

[276] Nawab Mehdi Ali, Mohsin-ul-Mulk, 1837-1907, served the Nizam in various capacities from 1874-1893, including Political and Financial Secretary. After his service to the Nizam he joined forces with Syed Ahmad Khan at Aligarh to promote Muslim education and the Urdu language. He later became politically active in the Muslim League. Webb later wrote of him: "… [Mehdi Ali] is one of the most intelligent and well-informed Moslems of India. He is eminently fitted to be a leader, but prefers to labor for Islam and Islamic unity on a perfect equality with all Mussulmans." "News Notes," *Moslem World and Voice of Islam*, January 1, 1896, p. 3; and Saeed, *Muslim India (1857-1947)*, 232-33.

gles—two mounted soldiers ahead. Soon afterward came a nobleman also dressed like a stage character. Saw many curious sights. Waiting to see marriage procession of the Nizam's sister with her cousin, a native nobleman Nawab Khurshed -ul-mulk Imam Jung Bahadur son of Sir Khurshed Jah.[277] Found a good place in a window of a small shop up stairs. Left Mehdi Ali there and went to the Mecca mosque for Magrib prayers—an immense mosque vast in its proportions.[278] After prayers returned to the shop and found Mehdi Ali in the carriage ready to go—said route of procession had been changed. Drove around into another street to the house of the Prime Minister where we found a most excellent place to view the procession from. Long covered corridor built expressly to see the Mohrum procession from. Eastern procrastination and dawdling governed the procession, of course. It should have started at 10 a.m. but really began to move at 8 p.m. The greater part of the interval people were waiting in windows to see the display and those who were to take part in it waited in the streets. These little sixpenny royal families have very little regard for the people who look upon the Nizam as a great person. When England has gobbled up his territory and banished him, they will not do him so much homage. We waited only about two hours, but we had easy chairs and a very palatable luncheon. When the procession finally came along it repaid us for waiting. First lot of men running with spears and shouting. Long lines of mounted native sol-

---

[277] The Nizam's sister was Najib un-nisa Begum, daughter of the Nizam Asaf Jah V. *The Chronology of Modern Hyderabad 1720-1890*, genealogical pull-out for "The family of Afzaluddaula Bahadur." She married Faiz ud-din Khan Bahadur, Imam Jung, Khurshid ud-Daula, Khurshid ul-Mulk, 1856-1925. Ibid., genealogical pull-out for "M. Rasheedud Din Khan." Sir Khurshid Jah, Shams-ul-Umara, Amir-i-Kabir, 1841- 1902, Commander of the Nizam's Household troops and former member of the Council of Regency and Council of State for the Nizam. Buckland, *Dictionary of Indian Biography*, 235; and Lethbridge, *Golden Book of India*, 253.
[278] Accommodating up to 10,000 worshipers, construction on the Mecca mosque began in 1614 under Sultan Mohammed Qutub Shah (ruled 1612-1626) and was completed by Aurangzeb (ruled 1658-1707) in 1687. K. Krishnaswamy Mudiraj, *Pictorial Hyderabad*, 12.

diers—elephants in gay trappings—a hundred men bearing tin foil figures of horses, towers, flowers &c—native bands—pipes and tom-toms—more soldiers—groom covered with glittering armor of spangles,—horse also decorated—bride in gorgeous palanquin—relatives on foot and on the backs of a herd of splendidly caparisoned elephants—bullock cart nicely spangled, bullocks horns and heads decorated with gold bands and red paint. One elephants head frescoed in bright colors and his tusks having gold bands. Other bullock carts—tinsel or tin foil objects, troops of foot-soldiers with long spears decorated with flowing ribbons—elephants with howdahs on their backs. Altogether a grand picture of old oriental life. Street was dark and very few torches so could not see very well. Hyderabad is full of interest to me and I would like to stay here a long time if I could only go about on foot. Returned home and found some friends at dinner. Got into bed a little after 10. Coming home two native soldiers on camels followed carriage.

November 22, 1892

Arose at 4:30,—a little warmer—had to put on thinner clothes. An hour later so cold I had to put on thick ones. Spent most of the morning writing on my lecture for Friday without interruption. At 5 p.m. started out for a drive and went to the large tank that supplies the city with water—fine drive along side of tank with excellent view of the surrounding country. Englishman rowing with girl in end of boat steering. Drove through the Nizam's public garden and past old grave yard. Evidences of the dim and misty past. Went to the house and private mosque of Shah Abdur Rahim who is sort of high priest to H.H. the Nizam.[279] Queer old-fashioned oriental house with a deep well in the centre and niches around its walls where one can sit and keep cool when the water isn't too high. The whole place reminded me of a cave

---

[279] Syed Shah Abdur Rahim Saheb Kumaisi-ul-Kadri, came to Hyderabad around 1863 and succeeded his uncle as the Nizam's spiritual advisor in 1878. A. Claude Campbell, *Glimpses of the Nizam's Dominions*, 329.

in the solid rock. It had a musty smell and a dismal gloomy look. We chatted for half an hour and were offered betel-nut which I refused. Then my hands were anointed with sandalwood oil. I was treated very courteously the Shah taking me by the hand and leading me to the door his bright little son holding his left hand. Two boys preceeded me with lighted candles and I was escorted to the carriage with a great deal of ceremony. Arriving at the club we found several persons waiting for us. Later on Mohammed Sharful Hak, of Delhi, a very learned Mohammedan missionary called and his visit proved very interesting.[280] He is only 31 years old and is proficient in Greek, and Arabic besides Hindustanee, Urdu and some Sanscrit and English. He has been a careful and earnest student of the Bible and has detected many of the interpolations and false translations in it. He also read the Koran in a very musical and attractive manner. A young M.D. and Theosophist was also present and talked.

November 23, 1892

Arose at 5:30 a little late. Spent the morning writing letters. The little son of Shah Abdur Rahim took tiffin with us and I gave the little fellow two Jap. silk handkerchiefs.[281] He is 9 years old and looks more like a girl than a boy. He is quite pretty. After tiffin I spent most of the p.m. looking over a bound volume of the London "Punch." What stupid brutes the English must be to call such an inane vapid publication humorous; and yet I am told they laugh ready to split their sides over the flat and pointless stuff it contains. About 4 we began to get ready to go to Secunderabad, three miles distant, to get some under drawers for me. Hassan Ali's slow and listless movements kept us waiting half an hour. Then Hajee Abdulla and I went to the carriage and waited for

---

[280] Mohammed Sharful Haq was the subject of a book entitled *Al-Bashul-Jaleel, or The Delhi Controversy between Moulvi Sharful-Haq Kadri Jalali and Rev. M.G. Goldsmith*, advertised in most issues of the *Moslem World and Voice of Islam* in 1895.

[281] Shah Abdur Rahim had three sons Abdul Hye, Ghulam Nuruddin, and Abdul Razzak. Campbell, *Glimpses of the Nizam's Dominions*, 329.

some time but Hassan Ali did not appear. I went to his room and found him squatted complacently on the floor preparing "pan supari"or betel nut. I aroused him and he languidly arose and came to the carriage after two or three stops on the way. Like most Eastern people he hasn't the remotest idea of the value of time. We had a very pleasant drive to Secunderabad obtaining some very pretty views of the country on the way. We found Secunderabad a dirty, dusty place. The only interesting feature of it was its oriental appearance. I wanted to stroll up one of its narrow streets to see native life, but we fooled away so much time that I was deprived of that pleasure. We went to one Mohammedan store to get the drawers but the proprietor didn't have what we wanted. He sent a young man out to look for them but he failed to find any and we had to go to another store where we found what we wanted and at a very reasonable price. We then went to a small mosque near by and said our Magrib prayers and started for home. At 6:30 went to house of Shah Abdur Rahim where we were invited to dine. Found quite a number of Mussulmans there and after a general chat for about an hour and a half we had a very generous dinner seated on the floor. There was a profusion of dishes, all well cooked and neatly served and with only a reasonable amount of pepper in them. I never saw so much pepper used as the people of this country put into their food. It is often so hot that I cannot eat it at all. After dinner we squatted on the floor in the corridor and then went into an adjoining room where Hassan Ali on behalf of Shah Abdur Rahim made a speech, addressed to me, thanking me for attending the dinner. I responded and spoke for about half an hour. Then the guests departed and Hajee Abdulla, Hasan Ali and myself were garlanded with flowers and were given little crystal vials containing oil of sandal wood. We were then escorted to the carriage given a cordial good-bye and went home. On arriving there found Mohammed Sharful Hak, the missionary waiting for us. He began to talk in Urdu to Hajee Abdulla and Hassan Ali and as I was sleepy and could not understand what they said I went to bed. The last thing I heard before going to sleep was Mohammed reading aloud the Koran in the musical Arabic.

November 24, 1892

Arose at 4:30 Nawab Mehdi Ali had promised to call at 8 and take us somewhere, but as he had not arrived at 8:30. Hajee Abdulla and myself went to the P.O. to mail some letters and from thence to Abid and Co's to get an empty box to send some superfluous clothes to Bombay. While we were there Hassan Ali came breathlessly in and said that the Nawab had been looking for me and was waiting outside in his carriage. I went out and getting into his carriage we drove through a very old portion of the town to the house of Nawab Server Jung, who is supposed to be one of the Nizam's right hand men.[282] Found Shah Abdur Rahim there with his son and several other Mohammedans. We had a very pleasant chat Jung talking English fluently and evincing no little intelligence and judgment. The object of the visit, I inferred was to get Jung to induce the Nizam to grant me an audience. These little six-penny monarchs are very difficult to approach except by good-looking actresses. After our chat we drove to the residence of Nawab Mehdi Ali, a beautiful mansion situated on rising ground and having from its upper windows a most charming view of the city and surrounding country. The grounds were finely laid out but badly kept and from a fountain in front a stone stairway led up on a terrace and thence to the house. The inside was furnished gorgeously in European style with elegant paintings and costly bric-a-brac and the purdah was modified by having a low screen across the room which separated male callers from female. The carpets, furniture and everything about the house were superb and showed that the owner was possessed of good taste as well as wealth. The house was a palace in its belongings and was much more elegant than the White House at Washington. While I was there a lady dress-maker from one of the shops was sent into the zenana to measure

---

[282] Nawab Agha Mirza Beg Khan Server Jung, 1849-?, a close advisor to the Nizam who had a hand in the educating the sovereign as a young boy. Campbell, *Glimpses of the Nizam's Dominions*, 365. For a further study of Agha Mirza Beg, see Agha Mirza Beg Khan, *My Life: Being the Autobiography of Nawab Server-ul-Mulk Bahadur*.

the Nawab's wife for a dress. This was the refined and elegant side of Mohammedan life. The other extreme shows up in the fanaticism of Bombay. Came home and found a Mussulman with his two little boys, had come to take tiffin with us. After tiffin packed box of superfluous clothing to send to Bombay and then wrote letter to Ella. About 3:30 a friend came in and, at my suggestion sent for a juggler. One came with two boys, one about 12 and the other about 16, or 18. The juggler was a thin, wiry fellow and like all of his kind, could talk at a rapid rate. His conversation with the older boy, while he was performing the tricks was very funny although I couldn't understand a word of it—it was the serious manner rather than the matter. His tricks were all good although very mechanical. The first was an mirage that shot a camel with a bow and arrow,—a very simple, but pleasing trick. Then he took a small metal bowl, set it on three stones, put some dirt in it and filled it with water—then put two little wooden ducks in it and while sitting at a distance of about 3 feet from the bowl he made the ducks bob their heads under water keep time to the squawking of his pipe and go completely under water. How he did it I could not understand. Then he performed several very commonplace tricks and finished with the basket trick in which the boy was hidden. We then went for a drive and visited the regatta, the first held in Hyderabad and by the English, of course. Saw the ladies' race—a sad and pitiable spectacle. Four women rowing a race before a mixed audience. Was put in the ladies stand and was introduced to the Primer Minister, a very affable monkey who has been bitten by the same snake that has bitten most of the more intelligent Indians: British customs and vices.[283] Why these poor, addle-pated people will sacrifice their manhood and play lick-spittle to a lot of beefy, whiskey soaked brutes who despise them I cannot imagine. They are the remnants of magnificent races but how they have degenerated. I was told to-day that the poor Nizam, a young man about 27 years of age often spent two weeks in the Zenana

---

[283] The Prime Minister was Sir Asman Jah, 1839-1898, who served in this position from 1887-1893 and was a prolific builder of palaces around the city. Buckland, *Dictionary of Indian Biography,* 18.

without being visible to anyone. He will probably share the fate of Thebaw of Burmah.[284] After we had wearied of the regatta we drove to the Public Garden where workman were putting up the tent for my lecture, under the direction of Shah Abdur Rahim. Said our Magrib prayer there and came home.

November 25, 1892

Was awakened at 2:30 by hearing rattling of the blinds on the other side of the house. I thought of the stories I had heard of cheetas, leopards, panthers and tigers who came into the compounds of residences and suspected that one of these animals was clawing at the blinds and trying to get in. It proved to be a cat which had been shut in the house and was trying to get out, but the noise robbed of more than an hour's sleep. Arose at 5 and prepared for my lecture in the Public Garden. At 7:30 Nawab Mehdi Ali called for me in his carriage and took me to the garden where a large crowd had assembled. The lecture had been announced for 8 o' clock, and as it was too early to begin, a native gentleman read a long poem in Persian, and by the time he had finished it was after 8 and the tent was filled. I suppose there must have been about 2,000 people present. I was introduced by a native gentleman who addressed the audience in Urdu for about five minutes and then I delivered my lecture speaking a little more than an hour. The audience seemed to be very well pleased. The lecture was translated into Urdu by Hassan Ali. One or two speeches were made by native gentlemen and when the proceedings were finished and I had passed through the usual hand-shaking ordeal we drove to the residence of Hafiz Ahmed Raza, one of the High Court Judges,

---

[284] King Thebaw, 1858-1916, of the Konbaung dynasty in upper Burma, had his kingdom annexed by the British in 1885 mainly in response to perceived French intentions to do the same. He died in exile in India. John F. Cady, *A History of Modern Burma*, 116-21.

where we had been invited to tiffin.[285] Found about a dozen persons present—among them the suspicious skeleton with the piccolo, hesitating voice—he had his leather leggings—possibly to keep the snakes from biting his shins. After the tiffin which was very good, we went to the Mecca mosque for prayer. Our carriage was one sent by the Nizam for our exclusive use, thus giving us two carriages; one for the morning and the other for the evening. The mosque is the largest I have yet seen and, I should think, will accommodate at least 15,000 people at prayer. After the zohar prayer there was a funeral service after the simple Mohammedan style and then Hassan Ali delivered a brief lecture. As it was in Urdu I occupied the time in looking at the nests of bees and hornets away up in the arches and the clouds of doves that circled about the grand edifice or lodged in the niches in the cornices and towers. The building and its worshippers were calculated to inspire one with awe. After the lecture I started hastily for the door to escape the crowds of hand shakers, but didn't succeed in escaping them all, for they blocked the way and it was only with great difficulty that I could make my way through them to our carriage around which they crowded in great numbers. We drove through the streets of the old city and saw the usual sights—among them a camel carrying a load of people—and out upon the stone bridge over the river, where two elephants were bathing and drinking. On our arrival home I took a nap, awaking just in time for the Asr prayer. After Magrib we went for a drive and returned about 7,—saw nothing of special interest. Went to bed about 9:30.

November 26, 1892

Was partially awakened at 4 by the cannon but fell asleep again and did not awake until nearly 6 when the sun light was just beginning to tinge the east. Hurried to awaken Hajee Abdulla and Hassan Ali—found that there was no water for

---

[285] Hafiz Syed Ahmed Raza Khan, 1841-?, held several judicial positions in the Nizam's courts. Campbell, *Glimpses of the Nizam's Dominions*, 80.

bath. Said prayers and had bath afterward. Received, after breakfast, a note from "G. Hanna" asking for the privilege of an interview.[286] Supposing that he was a truth-seeker I replied telling him to call at any time. He called about 10 and proved to be a fanatical Christian missionary—a graduate from the Salvation Army. I gave him a very plain lecture and sent him away somewhat humbled. It is singular how tenaciously these professional Christians cling to their degrading superstitions. I hope I shall not be annoyed by any more of them. Hassan Ali went to Secunderabad to get my lecture printed. Received from Cheragh Ali a copy of his excellent "Exposition of the Jihad."[287] Spent most of the day in writing and reading. About 3:30 p.m. just as I had finished a short nap, the small Doctor came to accompany me to Golconda.[288] While we were taking tea it was discovered that one of the carriages was out of order and that there was only one at our disposal; this Hassan Ali wanted to use to take him to the Mecca Mosque where he had promised to lecture at 4. I decided to go with him to the old city and stroll through the streets to have a look at the shops and people. The Dr. and I left the carriage in one of the principal streets and it took Hassan Ali to the mosque, returning and remaining with us afterward. We inspected the various shops about us and were soon surrounded by the usual crowd. One enthusiastic Mussulman insisted upon my accepting two handsome pairs of slippers and I reluctantly did so. The Dr.

---

[286] G. Hanna was possibly an English railway superintendent. Blunt, *India Under Ripon*, 51.

[287] Cheragh Ali, 1844-1895, a confidant of Sir Ahmad Khan, held several positions in the government, including Assistant Revenue and Political Secretary in Hyderabad. Although his interpretation of Islam was rather liberal several of his works were well received in India and abroad. Singh, *Encyclopaedia of Muslim Biography*, 2:187. In his *A Critical Exposition of the Popular Jihad* (1885) Ali argued that the wars of the Prophet Muhammad were all defensive and thus jihad was only permissible as a defensive response. P. Hardy, *The Muslims of British India*, 113. This work was also reprinted serially in *The Moslem World* July-October 1893.

[288] Presumably Dr. Nishikant Chattopadhya, see below, November 27, 1892.

wanted me to share the plunder with him but I was too self-ish—I wanted both pairs myself. Visited the native silver shops and bought a small watch chain. Stopped to look at something and two fanatical Mussulmans accosted us. One claimed that I had offended the whole Mohammedan population of Hyderabad by not making proper genuflexions at prayer, and the other insisted that is was my duty to adopt the Mussulman dress at once lest I should be mistaken for a Kafir. Poor, benighted creatures! they have no more idea of the true spirit of Islam than the cows or horses. Ignorance and superstition have done a fearful work all over India. After a time we entered the carriage and drove along one of the oldest streets in which there were many things I would have liked to see, but the Dr. like all the other chaperones I have had in India would not tell the coachman to stop nor to drive slowly. He was bent upon getting to somebody's palace before sundown and I wanted to get back to the Mecca mosque in time for the Magrib prayer, as Hassan Ali was waiting for me. After several ineffectual attempts to induce the Dr. to stop the carriage I went at the driver myself and managed to make him understand that I wanted to turn around and go to the mosque. The Dr. finally consented, very reluctantly, to tell him to drive slowly. All the people I come in contact with seem to be in a conspiracy to prevent my seeing anything that interests me. They are always ready to show me some new building of English design or anything European in its general character but they seem to be determined that I shall not see Indian life and Indian curios. Saw a pair of jugglers on the street entertaining a crowd of natives and I wanted to see the performance and asked the Dr. to stop the carriage; but he sat there like the sphinx and never moved. In sheer desperation I punched the driver with my umbrella and managed to stop him. The show didn't amount to much but I saw it. Found the people praying at the mosque with Hassan Ali in the front line. After prayer I made a short speech and one Mussualman became so enthusiastic that he announced the willingness of himself and his brethren to lay down their lives for me if I wanted them to. The usual hand-shaking followed and a great crowd surrounded our carriage as I got into it. On the way home we met (or overtook) a marriage procession which had 7 gaily caparisoned elephants and a camel in it; also a band of native music—pipes and tom-toms—and a crowd of followers. Arrived at home I found that Cheragh Ali had called and left his card. Hassan Ali and myself went to his house and found him just going out to dinner. On our return to the club found invitation to take breakfast next day with the Prime Minister.

November 27, 1892

Arose at 4 and wrote until 8 when Cheragh Ali called and we had a general chat. He brought with him an old man who had made the Bible prophecies the study of his life. He couldn't speak English so we didn't get on together. He promised to write some Urdu matter for the paper. While we were talking Nawab Mehdi Ali came to escort me to the residence of the Prime Minister and after chatting awhile we started. Had a very pleasant drive over a portion of the country I had not seen before. Quite an oriental aspect. High land and fine, cool breeze. Curious old ruins and mud hamlets, brooks and river,—rocks and turreted castle. Passed summer house of Prime Minister a turreted stone building resembling castle.[289] Arrived at Soorunagar, his usual place of residence, we found dilapidated walls, an ancient fort and collections of mud, chunam and stone houses.[290] Plenty of dirt. But inside the inner wall all was different—a new house as white as snow, just finished. Workman were employed in large numbers about the grounds. We were ushered up a broad flight of steps into a corridor decorated with handsome vases of flowers and then into a reception room from which we could get a glimpse of the parlor and retiring rooms. The furniture seemed all as new as the house and all very rich and elegant—European style, of course—the poor fellow is not only obliged to ape English manners and customs but is com-

---

[289] Asman Jah built several palaces, however, Webb's description and its location fits that of Asman Ghar a palace designed by the Nizam. Campbell, *Glimpses of the Nizam's Dominions*, 217-18.

[290] Often spelled Saroornagar or Sarurnagar. For a further description of the palace see, Ibid., 220-21

pelled to keep an English spy in his house who drives with him and watches all his movements. The Nizam, I am told, is also subjected to the same humiliating treatment.[291] The native government exists only by permission of the English and the people have neither the strength nor the courage to rise and drive the beef-eating brutes out of the country. It is a sad and pitiable spectacle. The house seemed overstocked with paintings and expensive bric-a-brac but there was a sad lack of taste in their arrangements. Carpets, curtains &c all very elegant. There were about a dozen persons present, including the English supervisor, all prominent in the Nizam's government. I was given the place of honor at the right of the host a very pleasant sort of man but not a genius. He didn't seem to have the conversational gift at all but he could smile and smile and smile and say "yes" and "no". The dinner was excellent and well served. At its conclusion Hassan Ali barked for 15 or 20 minutes about the American mission and Mehdi Ali followed him in a short speech which seemed to highly please all present. They spoke in Urdu and I was, therefore, relived of the duty of responding. We then retired to the reception room and after inspecting the bric-a-brac &c took our leave. Spent the greater part of the afternoon as usual in reading, writing and sleeping, and about 5 started out for a drive. Went to the public garden and saw figures representing the aboriginal races of India and also inspected the zoological collection. The latter is small and not well caged. There are three remarkably fine royal Bengal tigers, and three good leopards. The collection of monkeys and birds is inferior. We next called at the residence of Nawab Mehdi Ali where we said Magrib prayers and then talked about our departure from Hyderabad. He said that the Prime

---

[291] The British used Residents to watch over the Nizam and his advisors. Their purpose was "for gaining the Nizam's good will and esteem, and, at the same time, to discover any intrigues that may be meditated" and to "to keep a watchful eye upon his Highness's conduct, and to endeavour by every means in their power to establish a confidential and friendly communication between the two governments." Mudiraj, *Pictorial Hyderabad*, 111. Sir Trevor Plowden, 1846-1905, was Resident in Hyderabad from 1891-1900. Ibid., 122.

Minister would write to the Nizam to-morrow and arrange for an interview and that we would probably get away on Saturday. Little Dr. Nishikant Chattopadhya, called in the evening as usual and he Hassan Ali had a talk. Hassan Ali usually monopolizes the talk. Of all the lazy worthless fellows I have ever met I think he is one of the most incorrigible. Got to bed about 10.

November 28, 1892

Arose at 4:15 and had a bath. Awakened Hajee Abdulla and Hassan Ali. The former got up, bathed and prepared for prayers; the latter got up, scratched himself and then went to bed again. Slovenly lazy and indolent to the last degree, his only virtue seems to be that he can talk and he does talk at a most vigorous rate whenever the opportunity offers. Soon after breakfast the Prophecy crank came in with a friend and proceeded to explain the 7th chap. of Daniel, little Dr. Nishikant acting as interpreter.[292] About 8 another friend arrived to take us out to see a dervish. We took our own carriage and drove to a little mosque on a hill about 4 miles from the city. It was a beautiful place and afforded us an extended view of the surrounding landscape—valleys and hills, and rocks and in the far distance a hill with what looked like a castle on its extreme summit. A fine breeze was blowing and the sun shining so that the prospect was very fine. In a little room or cell at the left of the picturesque little white mosque we found the dervish, an emaciated little man apparently about 60 years old and crouched upon a bench and wrapped in a cloak of coarse yellowish stuff. His marked cleanliness, both of person and clothing as well as of his cell struck me at once; he was very different in this respect from the Hindu yogis. His hands and face and beard were noticeably clean and well-kept and his eyes were clear and intelligent. It is said that he was formerly one of the judges of one of the local courts and was the disciple of a dervish who lives where he lives now. He was so abstemious and economical in his

---

[292] The 7th chapter of Daniel purportedly portends the rise and fall of several great empires.

habits that he was considered a miser. About 12 years ago his teacher died and he went to the funeral and never returned. He left wife children, property, everything and gave himself up to contemplation in the cell of his master. He said that I was born for the work before me and that it would result in great success. He prayed with us or for us and gave Hassan Ali some good advice which probably went in at one ear and out at the other. We drove directly home and I put in the rest of the day in reading. Dr. N. came in about 3.30 and bored me until I got up abruptly and left him.[293] About 4:30 Hajee Abdulla and Hassan Ali and myself went out driving. Went to a common where a number of natives were doing the English—tent-pegging.[294] There was quite a large crowd and a band, and a sort of series of athletics were going on. Some of the natives are quite up to the English in these beefy exercises. After driving through the Public Garden we went to the residence of Mehdi Ali where we had prayers and a chat, a number of Mussulmans coming in during the talk. Mehdi Ali had been reading Bob Ingersoll and was highly pleased with him.[295] When we came home we found three or four persons waiting who had come to dine with us, one of the prominent brethren, an ex-judge, who subsequently appeared at the table with his head bundled up in shawls, having made a dinner in my honor. The table was very handsomely set, my name "Mohammed Alexander Russell Webb" being traced out on each side in flower leaves and petals. The little man with the spindle-shanks and piccolo voice sat at my right and a large man with a bad breath, at my left. One of his front teeth, which had the appearance of a tusk, wobbled in a most uncomfortable manner. The dinner was a quiet affair and no speeches were made. After it had

---

[293] Presumably Dr. Nishikant Chattopadhya.

[294] Pegging: "An Indian cavalry sport, in which the player, riding at full speed, tries to transfix and carry off, on the point of his lance, a tent-peg fixed in the ground." *Oxford English Dictionary*, 2nd ed., s.v. "Pegging."

[295] Robert G. Ingersoll, 1833-1899, a lawyer and political campaigner. Traveled the country lecturing in support of agnosticism and railing against Christianity in the 1870s. His lectures were widely published. *American National Biography*, 11:649-51.

been eaten we all went into the reception room and the man with the tusk and ropy breath singled me out as his victim. For about half an hour he waggled his tusk and blew his breath at me and I had to endure it. I was glad to get to bed. Time 11.

November 29, 1892

Hajee Abdulla awakened me at 5 and we had prayers about 6. Hassan Ali was awake and kept awake until we had prayed. About 8 the Bible prophecy crank arrived and proceeded to elucidate the rest of the 7th chapter of Daniel. Had his interpreter with him but the little Dr. volunteered and did most of the work but for his own benefit and information without any reference to me. Old man said he knew all about Revelations but when he tried to explain the 12th verse of the 3rd chapter I saw that he didn't.[296] Tried several times to get his opinion of this verse but the little Dr. steered him off onto something else until I became exasperated and tried the other interpreter with no better result. Specimen of interpreting: while sitting at desk note arrives written in Urdu,—all glance at it and remark: "Invitation." I give it to the interpreter and after studying it carefully he lays it down with the remark: "Invitation." "Invitation to what?" I ask after waiting in vain for him to speak, "Oh, invitation to some priestly festival." "But where and when?" "At his house, Thursday." "Where house and what hour." "Why the man's house who wrote the invitation. He don't say what hour." And that was all I could get out of him. Then I had occasion to lecture the little Dr. who became too fresh and wanted to show that he had acquired the stupid English habit of guying. He apologized very profusely. The Bible prophecy crank showed his hand—he wanted to go to America. Told him I would communicate with him later on. After the symposium had broken

---

[296] "Him that overcometh will I make a pillar in the temple of my God, and he shall go no more out: and I will write upon him the name of my God, and the name of the city of my God, which is new Jerusalem, which cometh down out of heaven from my God: and I will write upon him my new name." Revelation 3:12 (KJV)

up I read a little and then took a nap. Had tiffin and went in carriage to see one of our friends who was sick. Found him suffering from a mild attack of dysentery. After brief visit returned home, prayed and I made notes from railway guide of our probable route. Letter came in Urdu asking me what sect I belonged to. The fanatics have commenced to manifest themselves—as usual, we have been here too long. Oh, how wretched and degraded they have become. If our Prophet could see them now in the flesh, how he would excoriate them. Read and had a brief nap until just before 4 when I heard a couple of bells ringing and surmising that Shah Ab- dur Rahim Saheb had sent one of his elephants for me I went to the door and, sure enough, there was one coming up the driveway, gaily caparisoned in red and having a bell hanging by a long chain on each side. As he strode along the bells would swing out alternately and the chains striking his side on the return would cause the bells to ring. On his back were three persons beside the mahout who straddled the beast's neck. After tea and prayers I went out and the elephant was drawn up alongside of the high stone abutment to the front steps so that I could readily step on its back. The pad was soft and comfortable, and small silver-plated knobs at each of the four corners gave us a place to hold on. When I had safely seated myself the huge animal strode off. The sensa- tion was awkward and jerky at first but after a time I became used to it, but my legs became weary from being cramped in an uncomfortable position or hung over the edge of the pad. Then, too, I had to keep my umbrella up, on account of the sun, and that tired my arm. We went through some of the suburban streets past old tombs and ruined mosques and shrines which looked as if they were hundreds of years old. One of my companions was the captain of a band of 100 Ar- abs and the other was a Mohammedan boy about 16 years old who seemed to claim social distinction because he could speak English and had been to Ceylon. His English was pretty bad but I easily understood him. He pointed out the various tombs as we passed along, and particularly directed my attention to that of a great saint. "He killed 700 men— kafirs—Oh he was a great saint." Passed a place where a market was being held, and saw about 2,000 natives gathered, and a fair representation of yogis, beggars and jugglers. A very picturesque sight, but the elephant wouldn't stop and let me enjoy it. Passed many old tombs and crossed very old bridge across the Moosa River supposed to be more than 100 yrs old. Hindu temple at its edge, almost wrecked, by floods—smaller temple further down. Women washing clothes. Old walls of the city—supposed to be twelve miles in circumference. Curious old wall with holes for guns. The scenery very picturesque and oriental. Passed plebian elephant carrying load of hay—smelled of our elephant to see if we were all right, much to my nervousness. Another elephant with red trappings and single rider of the road in a paddy-field. We went to Meer Alim's tank about 3 miles from club house—clear water—two steam tugs—open space within walls for Mohammedans to pray, capacity of 3,000 or more—used during Ramzan.[297] Walked on walls of tank—beautiful view—magnificent landscape—rolling country dotted with boulders resembling ruins,—turreted castles—mosques and shrines—cocoanut trees, paddy-fields &c. Tank made by building wall across valley. Carriage with Hajee Abdulla and Hassan Ali had followed us and I decided to ride home in that as my ride on the elephant had fatigued me considerably. The ride home pleasant but dusty. Found Moslem school teacher waiting for us to ascertain if I would allow him to take my photo. Little Dr. came in the evening to conciliate Hassan Ali who was offended because he imagined that the Dr. had spoken lightly of the Prophet in a lecture delivered two years ago. Poor, bigotted fanatic! A great fall made over a trifle and the Dr. in hot water. Got to be at 9:30.

November 30, 1892

Arose at 4:30. At 8 a.m. a brother from Secunderabad called to take me to see a very old saint. We drove through some very old streets in order to get there and hauled up at the gate

---

[297] Meer Alim's tank was a dammed lake nearly 17 miles in circumference. Its construction began under the Prime Minister, Meer Alim, ?-1808. Mudiraj, *Pictorial Hyderabad*, 193, 308.

of a very old mosque. After we had waited for about 10 minutes in the yard we were ushered into an inner yard where there was a portico with niches, many swords and daggers, and a carpet on the floor. We had not long to wait when the saint came out and many of his disciples who had gathered at this time greeted him cordially and kissed his hand. He is said to be 115 years old. He is certainly 80, and perhaps he is as old as his followers believe him to be. He is a fine looking old man with grey beard, clear eye and clean face and hands, although he is emaciated and his skin lies in folds on his hands. He also seems to be short of teeth. By the time we were well seated around him fully 50 persons had come in, many of whom were very hysterical. Hassan Ali talked with him and he showed that he knew at least the ethical principles of Islam. His followers seemed to worship him and hung eagerly upon his words, some of them occasionally breaking out into shouts and violent contortions. After an interview of about half an hour we started to go and our friend asked the old man when he could give me a seance. The latter thought it would be better to have it at once and when he made known his decision the more hysterical ones howled with delight. We then adjourned to the larger meeting place in the next yard where a scene was enacted something like that in a large spiritualistic seance in America. The old man threw his mantle over his head and dropped into a meditative mood after telling me to close my eyes and think of him. I kept my wits and think I discovered the secret of his longevity. After vainly trying to bring me under the influence he drew aside his mantle and indicated that the seance was at an end. While it was in progress several of the followers, screamed and went into convulsions showing all the evidences of the work of elementals. The old man expressed a desire to see me once more and in Arab costume, but I hardly think he will. Some of his followers are hopelessly mediumistic. On the way home we stopped and saw a herd of the Nizam's elephants and two tigers the latter used in training the elephants to hunt tigers. There were 40 in this herd,—he has about 60 and some of them were very fine beasts. One was mad and was kept securely chained. Water ran from his sheath constantly and from his eyes. He was said to be very fierce.

There was also a baby elephant and two or three with frescoed head and trunks. The two tigers were in excellent condition and seemed to be about 6 years old. Found several parties waiting to see me—one had come to stay all day but it got so chilly that he left within half an hour. We were to have gone to Shah Sahib's at 1 to go to Golconda but it was nearly 2 before Hassan Ali got sufficient life to start.[298] At Shah Sahib's we found several brethren who joined our party and 4 or 5 of them rode out on Shah Sahib's elephant. Golconda is 6 ½ miles from Hyderabad and the drive took us through some of the suburban streets where dirt and children reigned—little shops, yogis and beggars—out into the country past ruined mosques and shrines and evidences of past glory. Passed a large tank and entered the walls of Golconda through immense iron plated, spiked and knobbed gates. City completely walled. Evidences everywhere of former magnificence—ruined palaces and residences overgrown with rubbish—massive granite walls and arches smeared over with mud. Streets quiet and dusty—dirty little hovels built in the archways of old palaces. From the streets look up to the fortress and rocky masonry dotting the hill. Boulders utilized to help form walls, and masonry plastered over and among them. Golconda once a large and powerful kingdom of the Decaan and contains mausoleum of the ancient kings.[299] Famous diamonds not found here but at Partial near the Southeastern frontier of the Nizam's territory—cut and polished at Golconda. Climbed up the hill by streets now deserted which once resounded to the tread of thousands of busy feet—past ruins of stately buildings overgrown with rubbish—massive walls with the roots of trees growing between the stones and forcing them apart—archways walled up and old mosque nearly at top. On top ruined residence showing traces of former magnificence,—secret passages—

---

[298] Shah Sahib refers to Shah Abdur Rahim. Server-ul-Mulk states it is a title used out of respect for religious men. Khan, *My Life*, 155.

[299] The kingdom of Golconda was founded in the early sixteenth century by the Turco-Persian Qutb Shahs, and was later annexed by the Mughal Empire in 1687. John F. Richards, *Mughal Administration in Golconda*, 1.

from top grand view of the surrounding country for many miles—lakes, tanks, heaps of boulders with buildings that looked like castles on top—palace of Fullacknoma in the distance—at foot of hill ruins of the palace of the Nizam covering, apparently two acres and showing traces of former architectural grandeur—grass and trees growing in the courts and parlors. Boulders piled up on the extreme summit as reminders of the glacial period. Fortified everywhere. The grandest view of landscape imaginable. Old granary built 250 years ago by Sultan Somebody,—old caves with Hindu idols walled up with solid masonry.[300] Capt. of Nizam's troops—dapper young fellow with side whiskers and general appearance of a British tourist sent up by Governor to attend us—had aerated waters and cigars served in the ruined residence on top of hill.[301] Then went down to officers' headquarters and had tea, biscuits and prayed at mosque. Nizam's soldiers on drill—spearsmen and swordsmen. Nizam has 5,000 Soldiers in Hyderabad and the English 8,000 in Secunderabad. English with improved weapons and Nizam's men with old-fashioned ones. Nizam has little chance of holding out when the Britishers reach for his scalp. The drive home pleasant but dusty. Stopped at Shah Sahib's and said Magrib prayer and had brief chat. Learned that Mehdi Ali had postponed his lecture—Britishers wouldn't like it. Hassan Ali went to see about it and as usual was very reticent and important. A very aggravated case of swollen head. Had some plain talk at the dinner table, as Shah Sahib called and matter came up of our going. It appeared that chief of police had told Mehdi Ali that speeches already made had aroused Mohammedans and that they threatened to kill all the English, and that if he made speech would do it at his own risk.

---

[300] The granary is most likely the Ambar Khana, believed to have been built by Abdulla Qutub Shah in 1642. Campbell, *Glimpses of the Nizam's Dominions*, 241.

[301] Possibly Nawab Mahomed Ali Beg, Afsur-ud-Daula, 1852-?, who matches Webb's description and was a high-ranking officer in the Nizam's forces. *Cyclopedia of India*, 3:305. As well, Nawab Major Afsur Jung is mentioned as leading the Nizam's regular troops and the Golconda Brigade in 1891. *Hyderabad in 1890 and 1891*, 77.

Lecture was postponed indefinitely, of course. Got to bed about 10.

December 1, 1892

Arose at 4:20. Hassan Ali awoke, turned over, grunted and went to sleep again while Hajee Abdulla prepared for prayers. While we were praying he got up. About 8 two brethren called to put me through an examination—one inclined to be fanatical and the other quite liberal. Avoided argument and the meeting passed of pleasantly. Little Dr. called and wanted to be taken to America. Hajee Abdulla, Syed Abdul Ghani and Hassan Ali had a conference and went out to call on Shah Sahib. I passed the morning in reading and sleeping—the little Dr. came in and bored me for a time about going to America. The whole day rather dull. At 4:30 went to Secunderabad to get some H.S. rupees for which I had paid on the previous day. Pleasant ride alongside of the tank but rather dusty the rest of the way. Prayed at a neat little mosque opposite the house of Mahomed Akber—child 4 years old praying other children also.[302] In the evening banquet given at the club in my honor at which Belgrami presided—large T shaped table decorated with flowers and fruits—good menu—Belgrami proposed my health and I responded in brief speech which was translated by Hassan Ali.[303] Received invitation to breakfast at palace of Fullacknoma. Found Bombay Gazette Nov. 29 with editorial in it. Got to bed at 11.

---

[302] Mahomed Akber was paymaster for the N.G. S. [Nizam's Guaranteed State] Railway and later treasurer of an Islamic mission at Secunderabad. "A New Mission," *Moslem World*, September 1893, p. 7.

[303] Most likely Syed Husain Bilgrami, see above, November 21. However, at the time of Webb's visit Imad-ul Mulk's brother, Syed Ali Bilgrami, 1851-1911, was also in the employ of the Nizam's government as Secretary of the Department of Public Works, Railways and Mines. Singh, *Encyclopaedia of Muslim Biography*, 2:150-51.

December 2, 1892

Arose at 5. After breakfast we drove to the Public Garden where Hassan Ali delivered a very effective lecture in Urdu to an audience of about 1,000. A company of native soldiers had been sent to preserve the peace, a rumor of a contemplated revolt of the Mohammedans having been put in circulation. They were drawn up in line on one side of the tent and a number of native policemen did guard duty. After the lecture a native poet read an original effusion the text of which I did not understand, but I heard my name mentioned in it and was told that it was very complimentary. I was subjected to the usual handshaking and accompanied Shah Saheb out to my carriage. We drove to the Club and thence through the old city to Fallacknoma the palace of the Nawab and it is a palace indeed.[304] Marble floors and stairways carved wood and richly frescoed ceilings,—oil portraits of father, grandfather, son, Prime Minister & others[305]—fine statuettes on neuel posts,—crystal chandeliers and oil paintings—axminster carpets and silk, velvet and lace portières—European richness and grandeur everywhere—banquet hall with table for 120. Elegantly carved furniture—marble verandahs from which magnificent view of the city and surrounding country could be seen—billiard room &c. Invited to breakfast about 20 at table including the inevitable English spy. I was at left of the Nawab—his son on right—not a very bright boy but quite English. The Nawab, Mehdi Ali sat at my left. Excellent dinner. Hassan Ali made a speech and so did Mehdi Ali and the Nawab subscribed 5,000 rupees to the

---

[304] Fallacknoma (common variant spellings include Faluknama and Falaknuma) was the palace of Nawab Vikar-ul Umra, Muhammad Fazil-ud-Din Khan, 1856-1902, who was Prime Minister from 1894-1901. The extraordinarily brilliant palace cost 3.5 million rupees to build and was sold to the Nizam in 1897 to offset its exorbitant cost. Mudiraj, *Pictorial Hyderabad*, 14, 180.

[305] Vikar-ul Umra's father and grandfather were Muhammad Rashid-ud-din Khan, 1815-1881, and Muhammad Fakhr-ud-din Khan, 1781-1863, respectively. He had two sons: Nawab Sultan-ul-mulk Bahadur, 1875-1949, and Nawab Wali-ud-din Khan Bahadur, 1878-1935. *Chronology of Modern Hyderabad*, Index, 4, Glossary, 14.

fund. Inspected the house under guidance of the Nawab and then went home. Mehdi Ali insisted that we must stay one week more. Took nap and spent the rest of the day lazily. After dinner Hajee Abdulla and Hassan Ali went to see Shah Saheb and I went to bed at 9:15. Visit to native photographer.[306]

December 3, 1892

Immediately after breakfast Hajee Abdulla and I took the carriage and started for the house of the bow-legged judge who had had my Hyderabad lecture printed. Hassan Ali had come to the conclusion that he was sick and couldn't move about. At the judge's house we had a wait of about 15 before he was in trim to receive us—came in with turban on and fully dressed with a smile a yard long on his fat face. Gave me 125 copies of the lecture and got into our carriage with us to ride to the club. We left him at the club to console Hassan Ali who, by this time was out of bed squatted on the floor sucking his hookah with great gusto. Hajee Abdulla and I then went to the house of the secretary of the club—a member of the interminable Jung family who wears constantly a white glove on his right hand.[307] Had a talk with him about our seeing the Nizam and he was of the opinion that we would see him "soon." This may mean two days or two months. The delay is simply exasperating and I would terminate it at once but for Hajee Abdulla. Went to a store and got some wrapping paper, and went home and spent the greater part of the day in getting my lectures ready for the mail. About 4 p.m. Shah Sahib came with his son and escorted us

---

[306] Possibly Raja Lala Deen Dayal, 1844-1910, court photographer of the Nizam for more than two decades. He also owned and operated commercial studios in several Indian cities. Clark Worswick, ed., *Princely India*, 18-20.

[307] Mahdi Hasan, Fath Nawaz Jang, 1852-1904, Secretary of the Nizam Club from its inception until 1892. At times he had been Chief Justice the Nizam's High Court and Secretary of the Home Department. Omar Khalidi, *An Indian Passage to Europe*, (forthcoming); and *Hyderabad in 1890 and 1891*, 61. Webb misconstrues the common title "Jung" or Jang as a family name.

to the residence of an old man who was an invalid and who wanted to see me. He seemed to be favorably impressed. His house was a small low bungalow very dirty and disorderly. He lay in bed—or rather, sat up—and the shelves and tables around him were covered with piles of books. The absence of woman's care was noticeable. Phenomenal boy who can parrot the Koran and knows everything. Thin lad about 12 years old with calm thoughtful eyes and very dirty clothes. Passed through the garden of another member of the Jung family, but this particular Jung was not at home. Parted from Shah Saheb's carriage after driving through the Public Gardens, and came home. After Magrib went to the P.O. to post lectures, accompanied by small boy—latter very talkative—I talked in English and he in Urdu—got on very well although we didn't understand each other. I said "Ha, ha" and "Gee ha" and "Bisheck" and that encouraged him so that he chattered away at a great rate.[308] Returned home—dinner—prayers and an argument with Hassan Ali about figures of animals on cloth—our mayordormo had discovered that my Jap gown had birds on it. Bed at 9:35.

December 4, 1892

Arose at 4:15 and after breakfast took carriage with Hajee Abdulla and Hassan Ali and went to see Mehdi Ali. He was not at home so we went to Shah Saheb's and had a general talk, remaining about an hour and accomplishing nothing, as usual. The movement seems to lack a head and makes little or no progress. We have been here two weeks and are just about where we were when we arrived. I never saw such a slow, lifeless community. From Shah Saheb's we went again to Mehdi Ali's and found that he had not returned so we went to the Jung who is Secretary of the Nizam Club. He could give us no encouragement but said he would see Mehdi Ali and try to find out what was going on. We had a pleasant chat and later his son-in-law, a young Dr. came in

---

[308] Ha = yes; Gee ha (ji han) = certainly; Bisheck (bi-shakk) = undoubtedly.

and was consulted in regard to Hassan Ali's liver. The liver went on strike a day or two ago, so, Hassan Ali thinks, and has been behaving badly ever since. It will not work, so the Dr. says, and the worst part of it is that he doesn't know why it struck. Of he could send down a Pinkerton regiment or even a native soldier or two the thing might be brought to terms but as that is impossible they are thinking of giving it a narcotic or something to quiet it until they can put it in irons.[309] Hassan Ali seems to have entirely lost control of it and lets it run things in his organism without trying at all to curb it. After about half or three quarters of an hour's chat we came home to breakfast. Just as I had sat down to write in came a very enthusiastic Mussulman—the Bulbul of the Deccan—who declared that his heart was filled with joy at seeing me and who introduced two quiet, innocent looking Nawabs.[310] He then read a poem addressed to me in Urdu, and seemed to feel relived. He was a great talker and seemed to have an echo in his stomach. I felt relived when he left after a short call. While we were at breakfast a young man called and waited until we had finished. He also was over-joyed at seeing me and said I had won the gratitude of every Mussulman in India. His name was Md. Asghar Hossain, of Lucknow. The afternoon was very quiet and I had a nap of about two hours. About 4:30 the Secretary of the club called and informed me that nothing had been heard from the Ni-zam yet. Later on the spindle-shanked judge with the piccolo voice called and piped for half an hour. Hajee Abdulla went Secunderabad and on his return I took a short drive. After-ward Syed Ahmed the fat, and myself went for a walk by moonlight in the Public Garden.[311] After dinner Hajee Ab-dulla went to Shah Saheb's and at 9 I retired.

---

[309] Pinkerton: "Applied to the semi-official detective force originally or-ganized and controlled by Allan Pinkerton." *Oxford English Diction-ary*, 2nd ed., s.v. "Pinkerton."

[310] Bulbul of the Deccan was a court poet in Hyderabad who could ap-parently compose poetry in seven languages. Blunt, *India Under Ripon*, 198.

[311] Syed Ahmed Khan, 1817-1898, well-known educational and religious reformer who spent many years in the service of the government. He

December 5, 1892

Arose at 4:30 and after breakfast Hassan Ali and I went to the bow-legged judge's house to get some more copies of my lecture. He was dressed to see us this time and took us up stairs into his drawing room. Very prettily furnished and with an air of neatness about it. No gew gaws nor evidences of woman's handiwork. House finely situated, having view of surrounding country—castle, shaped, and built on rock. Went to Post Office which is the worst I ever saw anywhere. If one of our frontier offices was managed in that way the post-master would be lynched. Employes Hindus of the most ignorant, stupid type who seem to have little or no idea of this work. Man wanted register stamp—clerk didn't know what it was—finally hunted about and came up smiling with five stamps. I wanted to know how much the postage on two papers to America would be—clerk didn't know—consulted two others—they didn't know—one weighed paper in his hand and looked nice, but still perplexed—finally bright idea struck one of them—concluded to weigh papers on scale and consult Postal Guide—happy discovery—mystery solved. European politely treated, but natives, other than Hindus snubbed and insulted. Came home and put in the morning wrapping and addressing lectures. After tiffin two men called who gave us news of the situation—Nizam had decided to see me but no day fixed. Hajee Abdulla and Hassan Ali to call on Private Secretary tomorrow and tell him all about our plans.[312] Nizam to listen in adjoining room. Spent afternoon in writing. After 4 had short drive—visited Public Garden and then went to Shah Saheb's and had Magrib prayer. Visited shrine of Saint in rear of Shah Sahib's house. Took off our shoes to walk in dust and sand. Came home leaving Ha-

---

also founded the Muslim Anglo-Oriental College in Aligarh. Mehra, *Dictionary of Modern Indian History*, 707-10.

[312] Syed Husain Bilgrami (See above, November 21) was the long time private secretary to the Nizam. He was appointed in 1888 and was still in the position as late as 1900. Imadi, *Nawab Imad-ul-Mulk*, 136; and Lethbridge, *Golden Book of India*, 489.

jee Abdulla and Hassan Ali to attend meeting of committee. Sat on front stoop and watched the immense bats that flew to large tree for fruit—bats 2 feet long or more hung by feet and eat fruit. Hajee Abdulla and Hassan Ali returned shortly after 8 and we had dinner—went to bed at 10.

December 6, 1892

Arose at 4:20. About 7:15 Hajee Abdulla and Hassan Ali, dressed in their best started for the Nizam's palace to lecture him through the Secretary. I commenced work on my Madras lecture. Hajee Abdulla and Hassan Ali returned about 11 completely crestfallen—had arrived at the palace too late—had agreed to be there at 7:30, but had to stop on the way to pick up the poet—he was late, of course, everybody is out here, and when they got to the palace it was 8:30,— Secretary had gone in to the Nizam leaving word that he would see them "some other day." Very unfortunate but a good lesson if those concerned would only take it to themselves. Procrastination and utter unreliability are the curses of the people of this country—after laziness, hypocrisy and a few other curses. In the afternoon fresh young reporter, G.A. Fernandez, employed by the "Deccan Budget" called to get report of banquet given to me at club about a week ago.[313] Specimen of newspaper enterprize—very characteristic—no newspapers allowed to be published in Hyderabad—British resident doesn't like them—so they are printed in Bombay and sent here. About 5 went out for drive—went to P.O. and then along tank road, stopping at small mosque for prayer. Then went to Mehdi Ali's but he was out—came home to dinner and sent boy to find out when Mehdi Ali returned and let us know. About 8 boy returned and Hajee Abdulla and I started on foot—found Mehdi Ali who gave us little satisfaction—main desire seemed to be to have us stay until Friday to hear his lecture. Didn't seem certain now whether the Nizam would see me or not, but was very positive last Friday, at Falacknuma, and on his declaration we remained another

---

[313] G. Arnold Fernandez, later wrote the novel, *The Romance of a Zenana* (1900).

week. Seemed very anxious and ordered his amanuensis to write at once to Server Jung telling him that we could stay no longer than Friday—probably destroyed note after we left. Politicians are about the same the world over. Came home determined to stay till Friday but not to be hoodwinked any more. Got to bed at 9.

December 7, 1892

Arose at 4:30. Resumed work on my lecture; about 10 Mehdi Ali called and said that if we did not receive a call to go to the Nizam before night we might as well go to Madras to-morrow. I knew that the case was hopeless and began to pack. But Hassan Ali's customary stupidity cropped out again and he wanted to see Mehdi Ali himself. He had been declaring for several days that he wanted to go to Madras and now that the opportunity offered he didn't want to go. His assininity will ruin him some day. He finally compromised the matter by lugging Hajee Abdulla off to Shah Saheb's. The latter immediately hustled his private secretary out on the fruitless job of trying to get us called to the Nizam and said he would call and see us at 4 and let us know. He came at 5:30 and knew nothing more than he did in the morning. About 6:30 Mehdi Ali came and declared that it was useless to stay longer and it was decided that we should go to-morrow.

December 8, 1892

Arose at 4:25 and after bath went over to Hajee Abdulla's room to pray. Awoke Hassan Ali who promptly turned over and went to sleep again as usual. Finally insisted upon his getting up and catching the 7 o'clock train when he sleepily informed me that we were not going to Madras; that Shah Saheb had sent a note over at 11 o'clock last night saying that we must stay. I was hot in a minute and declared most positively that I would start by the 7 o'clock train no matter what happened. Hurried the boys off for a bullock-cart and to order the carriage hitched up, much to the dismay of Hajee Abdulla and Hassan Ali; the latter managed to get awake and

192

alive when he saw that I was determined to go. It appeared
that Shah Saheb wanted us to stay so that we could hear Mehdi Ali's lecture to-morrow and so that I could be properly
"seen off" by a large crowd of hand-shakers at the station on
Saturday morning. The absurdity and stupidity of the pro-
posal fairly exasperated me. While we were eating breakfast
a man came from Shah Saheb's telling us that he wanted us
to stop there on our way to the station. I found that Hassan
Ali was scheming to prevent my getting the train and had
given orders to delay the bullock-cart. I finally got it started
however, Hassan Ali saying that his trunks would go by another cart. He is a "slick nigger" and although, lazy and shiftless to the last degree has head enough to keep from starving.
He was teaching school in Calcutta at 100 rupees a month
and gave up his pact in order to serve God—great sacrifice!
Last night he received about 300 rupees for 2 ½ weeks work,
as a missionary, and admitted that he was doing better now
than when teaching school. He is a black fraud. When we
arrived at Shah Saheb's instead of being ready to receive us
we had to wait 20 minutes for him. In the meantime Hassan
Ali sneaked outside so as to avoid acting as interpreter.
When Shah Saheb finally appeared he sent for Hassan Ali
and a lively argument ensued. I reiterated my determination
to start for Madras and was then told that my trunk had been
sent back to the Club by Shah Saheb. Then I was fully determined to go in spite of everything as I saw that all except
Hajee Abdulla had been scheming to keep me. I hastily, and
as politely as I could, under the circumstances, said good-bye
to Shah Saheb and ran out to the carriage. Hajee Abdulla accompanied me to the station and saw me into a 1st class car,
or alleged 1st class. The Nizam's State Railway is a caution.
While we were waiting for the train to start up came Hassan
Ali and Shah Saheb, and the latter renewed his request for
me to stay; or rather, entered a strong protest against my going. The train started while he was talking and, much to my
relief I was soon away from Hassan Ali and Hyderabad. In
the car with me was the little Secretary of the Nizam Club.
He was as calm and lady-like as usual. He accompanied me
to Vikarabad, a small station, about two hour's ride from
Hyderabad. We were now out of the boulder region and the

country was apparently fertile and well cultivated. While we were passing a section of the country pretty well covered with stunted trees, I saw two very large monkeys chasing each other through the brush. When they saw the train they scampered off at a great rate. Some of the vistas were very attractive—a sort of rolling prairie. We arrived at Wadi about 1 and much to my disappointment I found there was no hotel there—what had been called a hotel was only a refreshment room—there wasn't a bed to rent in town—town consisted of about 50 mud and stone huts. A fresh young man at the station told me I could sleep in the lavatory which was a stone cell on the ground floor which had a mirror a wash-stand, a washbowl full of dirty water and a stone bath tub which had never been used, in it, but no bed. The place had been whitewashed, apparently, some months previously and the spatterings of lime and dirt had never been removed—besides it had a disagreeable odor and I did not receive favorably the suggestion that I could haul one of the public benches inside and sleep there. Had a very good tiffin in the refreshment room and was told by the waiter that if I could get the station master's permission I could sleep in the ladies' waiting-room. Presently the telegraph clerk,—a Mussulman—discovered me and told two or three other Mussulmans. After he had bored me awhile he said he would go but would see me again after the train had gone. I made up my mind that he wouldn't if I could prevent it—but he did, about 5 o'clock and, with the assistance of a friend began to pump me. I endured it for a while and then shut him up by telling him that I had come to Wadi to get rid of people who bored me at Hyderabad and that it made me very nervous to talk. He took the hint and finally slid off. I then took a walk about the town; or, rather, up and down the R.R. tracks, feeling very blue. I regretted my hasty and stubborn action in leaving Hyderabad and looked forward to a night on a hard wooden bench with anything but pleasant anticipations. When I went to the refreshment room to get dinner at 6:15 I asked the Supt. If I could haul a bench in there and sleep. He said, no, but that the station master would probably allow me to sleep in a first-class coach if I asked him. I remembered my unpleasant experience with this station master when we changed cars at Wadi about 3 weeks ago on our way to Hyderabad and I didn't like the prospect. However as the Supt. wanted to take me to him I went and found the young man very affable. He quite promptly consented and ordered a yard-man to put a light in a first-class coach and show me to it. Compared to a hard bench it was luxurious. I gave the yard-man and his assistant one rupee each and the latter brought my traps from the refreshment room. By 7 I was as comfortably fixed as anyone could have asked. Rolled up in my rug with my spring overcoat and gray undercoat for a pillow and turned in at 8.

December 9, 1892

Arose at 6:30 and found water in the next compartment for a good wash. Felt much refreshed and more cheerful. Yard master's assistant thoughtfully came with water for me with the chill taken off but did not need it. Air quite chilly and bracing. About 7:20 went to refreshment room for breakfast. Lounged about the station until 8:30 when the train arrived from Hyderabad bringing Hajee Abdulla and an important letter from Hassan Ali. Much to my gratification the latter remained at Hyderabad to bark at to-day's meeting at the public garden. He could not resist the chance to talk. Hajee Abdull also gave me the gratifying intelligence that he desired to go on to Bombay and attend to some business, and wanted to know if I objected to going on to Madras alone. Not a bit of it. We put in the greater part of the forenoon in settling out plans and in a visit to the hospital where we had a chat with the talky telegraph operator and the native Dr. in charge. They brought tea and insisted upon our drinking it notwithstanding the fact that we, or I, had just finished my breakfast. When I refused they said they wanted me to drink just a little and they would drink the rest out of my cup, they considered it an honor to drink after me. I drank about one-third and they passed the cup around each taking a sip of it. Then we returned to the station and I dozed on one of the benches watching the sparrows as they flitted and twittered about the stone structure over my head, while Hajee Abdulla took a bath in the Lavatory. Three naked children—jet black and stark naked, having nothing on them but a string tied about the waist—played under the broiling sun in the vacant lot opposite. They strolled over to the station finally and amused themselves there. About 1:30 the train for Bombay came in and Hajee Abdulla got into a 2nd class car. His train started just as the train for Madras came in. I had a very bad tiffin at a very good price and got into a 2nd class compartment to find myself quartered with a nuisance in the shape of a half-caste lair who talked, and talked, and talked. He proved to be most disagreeable—took out a revolver as soon as the train started and laid it on the seat—afraid of train robbers—had seen me in Hyderabad and knew that I was an American—supposed that all Americans carried revolvers and wanted to show me that he was progressive. He proved to be a very stupid, self-conceited old ass and a most gifted liar. He rode with me from 2 to 8 when he got off at Gunta-kal much to my relief. He proved himself a hog by trying to get a carpet that I wanted to buy and by urinating on the closet floor and spitting on the car floor. I supposed I was going to have the compartment to myself for the night, but hardly had this pig alighted when a great fat Hindoo, weighing about 200 lbs came puffing in with his bundle of bedding and other traps. In the next compartment was a Hindu family with 3 or 4 small boys who continually ran to the water-closet and messed things up generally. The situation was too much for me so I bought a place in a 1st class carriage—and a very good one too, by the way; in fact the best I have yet seen. I had the whole double compartment all to myself all night. Rather dusty but seats soft and upholstered with leather and water closet and washstand very nice and clean. Car very roomy and comfortable. Went to bed about 8:30 and slept fairly well.

VII.

MADRAS

December 10, 1892

Arose at 5 and had a really good, refreshing wash and general clean up. Passed through very fertile country—rice fields, tall cocoanut and fruit trees,—rocky heaps to relieve landscape—shattered rocks—tanks, ponds &c a much better looking country than the Nizam's arid territory. At Arkonam at 5:30 had a very good cup of tea, ½ doz oranges and the first fresh grapes—except the Malagas—I have seen since I left America. Very good and very cheap. Also had buttered toast and some crackers—a very good breakfast. Enjoyed the scenery very much as it was a far better country than I had seen yet. At Perambur, the first station from Madras was met by a small delegation of Mohammedans who had prepared a good breakfast for me which I could not eat. At Madras the usual crowd was at the station to receive me and the usual hand-shaking process was gone through with. Was taken in the carriage of the President of Anjuman Islam to a handsome bungalow, where, after a refreshing bath I chatted with a number of Mussulmans until 11:30 when breakfast was announced. Very glad to see me and very much disappointed because I could stay only 3 days. Had made arrangements for me to lecture on Monday but as I could not stay arrangements were changed. Apparently very liberal people. At 1:30 started for Adyar. The drive there very beautiful through cocoanut groves—over level red, very wide roads—green paddy fields—bridge across the Adyar river built in 1845,

massive stone—beautiful view up and down river with
glimpse of ocean in the distance. Entrance to T.S. headquarters through grove of larch and mango trees—I think they
were, don't know positively—cool shade—grand old stone
and chunam house with assembly hall in front—found Col.
Olcott and the two Keightleys there—Dr. A. Keightley came
in 1st, then Bertram and then Col. Olcott[314]—chat in assembly room and then had tiffin, in detached building—very
good tiffin well served—everything looked as if the inhabitants were comfortable—Shrine room—took off shoes—two
paintings done in London of Mahatmas[315]—said to have
been precipitated by H.P.B.[316]—Precipitation of Stainton
Moses by H.P.B.—old portrait of Paraceleus and precipitation of face and head of adept by H.P.B.[317]—Grand photo of

---

[314] Henry Steel Olcott, 1832-1907, served as a Colonel during the American Civil War, and was a newspaper editor and attorney before joining forces with Madame H.P. Blavatsky to create the Theosophical
Society in 1875. Berger and Berger, *Encyclopedia of Parapsychology
and Psychical Research*, 298. Dr. Archibald Keightley, 1859-1930, a
physician, and his uncle Bertram Keightley, 1860-1945, a lawyer, became associated with the London Theosophical movement in 1884.
They were later the primary caretakers of the movement's co-
founder, Helena Petrovna Blavatsky. Meade, *Madame Blavatsky*,
382-86.

[315] "Exalted beings who, having attained to the mastery over their lower
principles are thus living unimpeded by the "man of flesh," and are in
possession of knowledge and power commensurate with the stage
they have reached in their spiritual evolution." H.P. Blavatsky, *The
Theosophical Glossary*, 201.

[316] Helena Petrovna Blavatsky, 1831-1891, Russian-born mystic who
preferred using her initials HPB. She traveled around the world studying various aspects of mysticism and spiritualism and in 1875 co-
founded (with Henry Steel Olcott) the Theosophical Society. Berger
and Berger, *Encyclopedia of Parapsychology and Psychical Research*, 38-39. For more information concerning Blavatsky, see Cranston, *HPB*.

[317] Should be "Paracelsus." Philippus Auraelus Theophrastus Bombastus
von Hohenheim, Paracelsus, 1493-1541, a native of Switzerland and
physician by training. Traveled the known world and studied alchemy
and occultism. Believed in the doctrine of Macrocosm and Micro-

H.P.B.—in fact photos of H.P.B. abounded—old shrine room torn away and new room built used by Dr. Keightley for a bedroom—Col. Olcott's room a cosy bachelor's den—books, cigars &c—asotea overlooking Adyar river,—very pleasant. Oriental library—shields—publishing office—Col. O. and Bertram had to go to R.R. Station to see some ladies off and left me with Dr. A. Keightley—had pleasant chat about T.S. matters and odd things generally. Returned to house, changed dress and after chat with prop'r of house and others on front stoop carriage called to take me to Umdah Bagh for dinner—residence of Mirza Ismail Khan Bahadur. Met Mrs. Mohammed Ahmed son of Mushtak Hosein,[318]—English woman, rather pretty, and her friend Miss Allen from England. Rather odd to see ladies among Mussulmans.[319] House situated in large garden—grand entrance—broad stairways right and left covered with red carpet—grand entrance guarded by soldiers. Nawab showed me number of mechanical toys French make[320]—had dinner with Nawab 2 ladies and 2 sons of Nawab, and Mohammed Ahmed. After dinner was introduced to a native king who came with a retinue of soldiers. His name was, Nawab Mir Fateh Ali C.S.I. of Banganapalli.[321] After a short wait went

---

cosm, relating the human body to the greater universe. Nevill Drury, *Dictionary of Mysticism and Esoteric Traditions*, 238. Adept: "In Occultism, one who has reached the stage of initiation, and become a master in the science of esoteric philosophy." Blavatsky, *Theosophical Glossary*, 6.

[318] Mohammed Ahmed, 1868?-1896, barrister and eldest son of Mushtaq Hussain, Viqar-ul-Mulk, 1841-1917, Judicial Inspector for the Nizam. Singh, *Encyclopaedia of Muslim Biography*, 5: 401-06. Webb later wrote of Ahmed, stating: "He is one of the rising young Mussulmans of India…His great mistake was in not coming to America to establish himself, where talent is fully appreciated." "News Notes," *Moslem World*, July 1893, p. 9.

[319] Possibly the same Miss Allen who worked in Hyderabad leading a club for non-Christian students and a Muslim Ladies Association. Raj, *Medievalism to Modernism*, 274.

[320] Presumably the Nawab is Mirza Ismail Khan Bahadur.

[321] Nawab Fateh Ali Khan, 1848-1905, succeeded his uncle in 1868 and ruled the Native State which was under the control of the governor of

to another room where dinner was spread on the floor Mohammedan fashion. Was asked to eat but could not—squatted for a while and that seemed to satisfy them. Nautch girls in parlor, men with fiddles, pipes and tom-toms—and brass clappers. Mohammedan girl sang and postured then very buxom and pretty Hindu girl loaded with jewelry and flashily dressed danced until I was weary. Men sang pipes squawked and tom-toms beat all the time in a very monotonous way. Nawab's son had diamond watch chain and diamond cluster ring. Got tired of the dancing and started to go—Nawab hung gold cloth and stone garland around neck and gave me pretty cut-glass bottle of altar of roses. Dancing girl had head dress of roses made like Davy Crockett trapper's cap—diamond nose rings. Got to bed about ½ past 10 and was devoured nearly all night by mosquitos.

December 11, 1892

Arose at 5:30 feeling rather old. Finally made boys understand that I didn't want them any longer and they went out. Got fairly good bath and good breakfast. Wrote to Ella and Shah Saheb and then prepared to lie down and take a nap. Hassan Ali put in an appearance about 9 as greasy and complacent as ever. He's a good natured idiot after all. I hardly think he means any harm—he's simply lazy and no good generally. After I had seen him I lay down and had a fairly good nap of about an hour and a half. Got up to prepare for tiffin expecting that I would have to go out to Aziz Bagh, where the tiffin was to be given at 11 but found that the house I was in was called Aziz Bagh and that the affair was to be very quiet—only the old fossils who composed the committee of arrangements. It was about the poorest and worst served of any formal meal I have had in India. The committee meant well but didn't know. Crows very numerous here and very cunning—when floor cloths were laid out to air they tried to carry them away. These crows would steal a hot stove if they could carry it off. After tiffin sat and chat-

---

Madras. He was removed from power in 1905 for alleged misgovernment. *Imperial Gazetteer of India*, 6:374.

ted awhile with several members of the committee, tried to sleep but couldn't and at 2:30 got up to prepare for my lecture which was announced to take place at 3. At 2:45 found no one around but the workmen who were engaged in putting up an awning in front of the house. About this time a rather beefy reporter for the Madras <u>Mail</u> came up and sat himself down on a sofa. Reminded me of some of our cheap reporters who smell around at public assemblages and don't know how to write up an item after they get it. Presently Col. Olcott and "Sydney" drove up. "Sydney" is rather slim and fresh and quite English. I'm afraid he is N.G. The Col. is rather disappointing. Thought him more sedate, serious and dignified and not given to worldliness. When he said his law practice in New York used to yield him a thousand dollars a month strange thoughts came into my mind. I'm afraid he is not all a disinterested worker for good. It was five minutes past 4 when I commenced my lecture and there were about 250 people present. Hassan Ali translated part of my lecture and then Col. Olcott spoke briefly but very complimentary of my address. He said that I was: "A devilish good speaker." This was said to me immediately after I had finished my lecture. We then had Magrib prayer and then Mohammed Ahmed, son of Mushtak Hosein made a brief address to which I responded briefly. I then went out to drive in an elegant turnout with the Turkish Consul, a merchant, who took me to the Sea shore, a fine drive and promenade where a band plays twice a week. Madras is far ahead of any Indian city I have yet seen. Wide, well macadamed streets, plenty of shade and good sea air make it very attractive as a place of residence. On our return an address was read to me from the Anjuman-Islami to which I responded briefly. I was told that it would be engrossed and presented to me at the station. Then we had dinner on the floor after I had eaten something at a table. The people seemed to think it a compliment for me to squat down with them. There were about 200 at dinner. After dinner chatted a while and was preparing to go to bed when Hassan Ali came to my room and wanted to introduce me to a Mohammedan saint. Went out and found a very tall man of the Hajee Abdulla stamp and was impressed with the idea that he was very good. As he could not speak English the interview was not very interesting. Started to bed and found that no water had been left for my bath—jar had been taken away. Started 4 boys in succession to work it up but was finally told that there was no water and that the water coolie had gone home. There were about 40 men and boys about the establishment and none of then could bring water. Insisted upon having it and it was brought while I went to bed. Retired about 10:30.

December 12, 1892

Arose at 4:45 and after bath &c went out on the front place to walk. As soon as it was light enough I began to write in my journal and after finishing tried to take a nap until breakfast time. Was awake when the boy said tea was ready. Found the tea there but no milk, teaspoon nor bread. One of the monkeys came in with his property smile and as soon as it dawned on him that I wanted bread &c, he started off and they were soon brought to me. Went to my cracker box where I had left two oranges and found that one of them had disappeared—niggers who slept in the wash-room must have nipped it. Made a fairly good breakfast and then lounged about waiting for the Turkish Consul to come and take me to the services of the opening of the new asylum for Mohammedan Converts at which ceremony I was to preside. While we were waiting up came a Mohammedan fakir and in spite of the efforts of the porter he stripped himself to the waist and began to drive an exaggerated ice-pick into the top of his skull. He had two of these picks, the heavy ends of which were adorned with iron chains. The iron points were about 6 inches long and tapered from ¼ of an inch in diameter down to a sharp point. He placed the point, apparently at the juncture of the sutures of the occipital and frontal bones of the skull, and with the heavy end of the other "pick" drove it in until it stood up firmly, without his touching it. It appeared to penetrate about an inch and he removed it with some apparent difficulty. The servants tried to drive him away but I insisted upon investigating the case to see if the iron really entered the skull. I saw the hole in the scalp and a small quantity of blood in the hair. At first I did not see this—there was

202

apparently no hole but it afterwards appeared that I had not examined carefully enough and that as no blood had then appeared, and as the matted hair covered the hole I did not see it. The fellow then took some iron skewers about two feet long and tapering from ⅛ of an inch in diameter down to a sharp point. He had about ½ a dozen of these skewers which he threw down upon the pavement. Taking up one he put the point in his mouth and forced it through the flesh of his cheek—on the left side—then he passed another through his right cheek, and another through his tongue. Then passed one through the muscles of the neck. The skewers were rusty and the pain must have been intense but the fellow never winced. The performance was genuine, without a doubt. When he had drawn out the skewers, I again examined the top of his head and the blood in the hair guided me to the hole. How the fellow endured this I can only surmise. I gave him a rupee and a companion of his wanted also to perform the feat. I begged to be excused. The carriage now came for us and we went to the ceremony of opening the new asylum for Mohammedan converts. I was selected to preside which I did with my customary grace and dignity. One of the officers of the Anjuman-Islami read a speech in which he complimented me very highly and which was translated into American by Hassan Ali. I then spoke about 15 minutes and my remarks were also translated by Hassan Ali. Then I had to run the gauntlet of hand-shaking. There were about 2,000 persons present and the job was not a pleasant one. We then went to the house of the Turkish Consul nearby and had a most excellent breakfast. There was a very large gathering present and the majority ate down stairs on the floor while I, three Englishmen, Mirza M'd Ismail Khan and two or three others ate at a table in a nice breezy room in the top story of the house. A Hindu photographer then took a group with me in the centre and we started for home calling on a very old and alleged wise Mussulman who congratulated me and wished me success. I packed up and lay down to take a short nap. Just as I arose a servant announced a visitor—it was Mohammed Ahmad. I hastily finished packing, gathered up my grip-sack and small things, and leaving my trunk to be sent on to the station by Hassan Ali. Went in Mohammed

Ahmad's carriage to his handsome home situated in a large and well-shaded compound. Found his wife gorgeously dressed in canary colored silk en <u>train</u>, and a profusion of diamonds and jewelry. Miss Kate Allen, her friend, was dressed in simple white. We sat in the front reception space and presently Mirza M'd Ismail came with three of his children. We had a very pleasant chat and a most excellently served tea at 4—it reminded me of home. Everything was delicate and in good taste. At 5 we started for the station, Mrs. Ahmad giving me a box of biscuits for luncheon. At the station found a large number of Mussulmans waiting to say good-bye. The Anjumani-Islam very generously paid my fare, 1ˢᵗ class to Agra, 90R's—10A's and gave me 45 rupees for feed on the way. This very liberal treatment was quite unexpected. In fact I had nothing to complain of at Madras except the mosquitos and the hard bed and feed at Aziz Bagh. At the station a fresh young native from the "Mail" attempted to interview me but his questions were so absurdly foolish that I had to shut him off, much to his evident disgust. The good-byes were affectionately said and four Mussulmans, the Turkish Consul among the number accompanied me to the 1ˢᵗ station out, Perambur, where they left me to proceed on alone. Found the carriage the best and most comfortable I had yet met with. Bought some milk at Arkonam which with the bread and biscuits I had, made a very good supper. Went to bed at 8:30.

VIII.

AGRA

December 13, 1892

Awoke at 5:30—had a good wash and made good breakfast out of milk, bread, biscuits, oranges and bananas. Passed through boulder district again—scenery very interesting—deer racing with train and bounded off toward heap of boulders—rugs at Adoni—Crossing the Tungabhadra River, grand view up stream—about 250 yards wide and clear as crystal, meandering in foaming streams among jagged rocks. At Raichur, at 10:45 had tea, toast and biscuits. Hindu man and wife in full English costumes—very funny. Crossed the Chicksur River at 11:45—broad stream resembling the Tungabhadra—women on rocks far below washing clothes—rock dam built partially across over which water washed and foamed very prettily—at Krishna swell native policeman with side whiskers brushed the wrong way and heavy moustache in the same fix—gravely saluted train as it arrived and departed. Knew that I was again in the Nizam's dominions a rocky waste of land at this season. Approached Wadi Junction with some trepidation—when went down, in kindness of heart and appreciation of favor of 1st class car asked station master if he wanted anything in Madras that I could bring up for him. "Oh, no, bring anything you like" Such gall was truly British and I didn't know how to get out of the fix. He came to the car while the talky telegraph operator and the Dr. were there and just faced it out. The other side of Shahabad, slate fields used for ballast Great Indian Peninsula R.R.—

approaching Kulbarga gathering of natives in field 4 camels—picturesque.[322] Picturesque group of 6 men and children 1 woman riding cow. Kulbarga in the Nizam's dominions used to be his favorite resort. Approaching station domes of white mosques in distance and apparently large city. At 10 p.m. changed cars at Dhond and had a very comfortable sleep.

December 14, 1892

Awoke at 5:30 to find myself at Monmar, where I had to change cars again. If I hadn't awakened myself I would have been carried on—such stupid arrangements, for R.R.'s ever saw. Met here L.C. Watkowsky, who used to live in Chicago—been here in Gov't employ 16 years—rode all day with him in 1st class coach and forenoon also with F. Armfield of Archer & Co. Am. book-makers. Dusty ride all day through fertile country well grown with trees. At Itarsi got very good through 1st class car for Agra and Witkowsky who had 2nd class ticket bribed the guard and got in with me. Nothing of importance to record.

December 15, 1892

Awoke about 5:30 and washed in water cold enough to make my finders ache; however, it was very refreshing and the atmosphere had quite a frosty suggestion and was very bracing. We rode through a country that had been worn into deep ravines and gullies by the action of water and presented a very peculiar appearance. It seemed to be worthless for the purpose of cultivation but would afford a most excellent hiding place for an army. Boy sold us two cups of tea at Dholpur—regular price 2 annas per cup—cups very old, unmatched and dirty looking—boy said 3 annas each—"why?" "2 anna 2nd kullass 3 anna first kullass." Witkowsky put him out with 4 annas. Arrived at Agra was surprised to see near station flocks of kites, hawks, crows and vultures perched on walls—swarms of them—the city presented very gray, dusty

---

[322] Should be Gulbarga.

appearance. Went to Laurie's Hotel and was given good room with bath attached; called for mosquito bar and was told that one would be given me soon; later on was assured that there were no mosquitos in Agra. Madras lesson no effect—accepted statement and did not insist on having bar. Hotel was formerly Gen. Husain Ali Khan's house.[323] After bath and breakfast, started out about 11 to visit the fort of the Emperor Akbar.[324] I had a very comfortable gharry and a Hindu guide who spoke very good English and seemed, at first, quite a treasure. But as soon as he saw me taking notes he was seized with the idea that I intended to write a book, and with an intense longing to have his name appear in that book. When I assured him that his name would appear he became consumed by a desire to have me put down regularly, and in his own language everything he told me. He turned out to be quite an ass. Fort is 1 ½ miles in circumference, of red sandstone and was built by the Emperor Akbar the Great in 1556. Entered by the Delhi gate—very massive and grand walls 70 feet high, surrounded by a moat 30 feet deep and about 20 feet wide. Next entered the Elephant gates or Jamul Futtah, in inner wall.[325] Formerly were two massive elephants, one on either side of the gate, but they were knocked down by Aurengzeb a grandson of the Emperor Akbar and about as great a fanatic as was produced in those days.[326] The immense gates covered with brass and copper sheets—guard rooms inside with square holes for the guards to peep out of and pop over any intruder. Elephant stables in front of gates now used as mess rooms for British troops and

---

[323] Husain Ali Khan,?-1720, Mughal viceroy of the Deccan (1715-18). Mehra, *Dictionary of Modern Indian History,* 640-43.

[324] Akbar the Great, 1542-1605, ruled the Mughal Empire for nearly half a century, achieving endless victories and creating the dominant power in the subcontinent. John F. Richards, *The Mughal Empire*, 56.

[325] Unclear reference, the Elephant Gate is known as Hathi Pol. E.B. Havell, *Handbook to Agra and the Taj, Sikandra, Patehpur-Sikri and the Neighbourhood*, 40.

[326] Aurangzeb, 1618-1707, like his grandfather, Akbar the Great, he ruled for half a century and implemented a strict interpretation of Islamic law, resulting in the ruin of many non-Muslim temples and buildings. Richards, *Mughal Empire*, 175.

recesses in walls by shoemakers who mend soldiers' shoes. Beautifully carved cornices of red sandstone. Walls chipped and cracked by weather. Passed, on the left the Danza's palace, a Marattah king but didn't...[327]

---

[327] Dansa Jat's palace was the seat of the Rajas of Bharatpur when they controlled the fort in the 18th century. Havell, *Handbook to Agra and the Taj, Sikandra, Patehpur-Sikri and the Neighbourhood*, 43. The second journal ends here, the whereabouts of succeeding journals is unknown. The last page had several rail stations and locations listed: Ahraura Road, Chunar, Pahara, Mirzapur

IX.

BOMBAY LECTURE—ISLAM

November 10, 1892

I have been requested to tell you why I, an American, born in a country which is nominally Christian, and reared under the drippings of an orthodox Presbyterian pulpit, came to adopt the faith of Islam as my guide in life. I might reply promptly and truthfully that I adopted this religion because I found, after protracted study, that it was the best and only system adapted to the spiritual needs of humanity; but this reply would be altogether too general to be satisfactory. Now, instead of telling you in detail how and why I became a Mussulman, I will try to tell you what Mohammed really taught, and endeavor to show that his religious system has a much more valid claim to the careful, unprejudiced attention of the Western world than the average Christian is willing to concede. It is manifestly impossible in a single lecture to present a full and complete exposition of the Islamic system—one that will satisfy the inquiring mind. I will, therefore, be compelled to generalize to some extent. And here let me say that I was not born, as some boys seem to be, with a fervently religious strain in my character. I will not even assert that I was a good boy, such as mothers sometimes point out as shining examples for their own sons. I attended the Presbyterian Sunday School of my native town, when I couldn't avoid it, and listened with weariness and impatience to the long, abstruse discourses of the minister, while I longed to get out into the glad sunshine and hear the more satisfying sermons preached by God Himself through the murmuring brooks, the gorgeous flowers and the joyous birds. When I reached the age of 20, and became practically my own master, I was so tired of the restraint and dulness of the Church that I wandered away from it, and never returned to it. As a boy I found nothing in orthodox Christianity calculated to win me to it, and in later years I encountered convincing evidences of its grave errors and insufficiency as a means of securing salvation, or of elevating and purifying the human character. Fortunately I was of an inquiring turn of mind; I wanted a reason for everything, and I found that neither laymen nor clergy could give me any rational explanation of their faith, but either told me that such things were mysteries or that they were beyond my comprehension. After trying in vain to find something in the Christian system to satisfy the longings of my soul and meet the demands of reason, I drifted into materialism, and for several years had no religion at all. About eleven years ago I became interested in the study of the Oriental religions, beginning with Buddhism as students of the Eastern system usually do. I will not weary you with details further than to say that at that time I had access to a most excellent library of about 13,000 volumes, from four to seven hours a day at my disposal, and that I was intensely in earnest in my efforts to solve the mysteries of life and death, and to know what relation the religious systems of the world bore to these mysteries. My mind was in a peculiarly receptive, yet exacting and analytical, condition, absolutely free from the prejudices of all creeds, and ready to absorb the truth, no matter where it might be found. Firmly materialistic, I looked at first to the advanced school of materialistic science and found that it was just as completely immersed in the darkness of ignorance as I was. It could tell me the name of every bone, muscle, nerve and organ of the human body, as well as its position and purpose or function; but it could not tell me the real difference between a live man and a dead one. It could tell me the name of every tree, plant and flower, and designate the species to which each belonged, and what were its apparent properties and attributes; but it could not tell me how and why the tree grew and the flower bloomed. It was absolutely certain that man was born, lived a

210

brief space and died, but whence he came or whither he went were riddles which they confessed themselves utterly unable to solve. "Those matters belong to the Church," said a scientist to me. "But the Church knows absolutely nothing of them," I replied. "Nor do I—nor does science," was the helpless, hopeless way in which he dismissed the question from the conversation.

I saw Mill and Locke, Kant, Hegel, Fichte, Huxley, and many other more or less learned writers discoursing with a great show of wisdom concerning protoplasm and protogen and monads, and yet not one of them could tell me what the soul was or what became of it after death.[328] "But no one can tell you that," I fancy I hear some one say. That is one of the greatest errors that poor, blind humanity ever made; there are many people who have solved this mystery, but they are not the materialistic followers of any creed. I have spoken so much of myself in order to show you that adoption of Islam was not the result of misguided sentiment, blind credulity, or sudden emotional impulse, but that it was born of earnest, honest, persistent, unprejudiced study and investigation, and an intense desire to know the truth. And now let us see what Islam really is, and you will readily understand, I think, why I have accepted it. If anyone should ask me to reply at once to the question: "What do the Mohammedans believe?" I would be as completely unable to reply, without hesitation, as I would be if asked: "What do the Christians believe?" The disposition of the Christians of all ages, from the days of Constantine to the present, to amplify and adorn their religion with ideas of their own has been shared by many of the professed followers of the Arabian Prophet, and there are prevalent among the Moslems of our time many points of belief which were never taught by Mohammed, and which are not entitled to a place in the true faith of Islam. The wonderful fecundity of the human mind is amply shown by the vast variety of conceptions—or misconceptions—which may be found among the religious doctrines of mankind, and

---

[328] John Stuart Mill, 1806-1873; John Locke, 1632-1704; Immanuel Kant, 1724-1804; Georg Wilhelm Friedrich Hegel, 1813-1863; Johann Gottlieb Fichte, 1762-1814; Thomas Henry Huxley, 1825-1895.

which were never a part of the original fundamental principles, but are the results of the biased speculations and misconceptions of those who have assumed to be religious authorities. It is a well-known fact that every one of the fifty odd different Christian sects bases its system of doctrines upon the Bible, and that the followers of each sect appeals to that much-abused book for proof that their belief is right and reasonable and logical, and that all the others are more or less erroneous. Wade through the tons of Mohammedan and Christian literature extant, if you have the patience, and go among the followers of the various sects and listen to their arguments, if you desire to acquire a perfect realization of the fertility of the human intellect and the elastic possibilities of religious literature. If you do not speedily find yourselves in a condition of doubt and perplexity closely bordering upon despair, you will have failed to follow the intricacies of the arguments put before you. If, from the multitudinous and multifarious conceptions you will encounter you can form a positive and satisfying opinion as to what Mohammed and Jesus really did or did not teach, you will do much better than anyone who has ever tried the experiment before you. There are certain clearly-defined dogmas which nearly, if not quite all Moslems believe in the abstract, but, with the possible exception of the Sufis, or esoteric Mohammedans, they are greatly divided in their conceptions of the Prophet's teachings.

Orthodox Mohammedanism may be divided into six heads: 1st—Faith in God, the one God, the Creator of all things, who always was and ever will be; the single, immutable, omniscient, omnipotent, all-merciful, eternal God. 2nd—Faith in angels, ethereal beings perfect in form and radiant in beauty; without sex, free from all gross or sensual passion and the appetites and infirmities of frail humanity. 3rd—Belief in the Koran as a book of divine revelation given at various times to Mohammed by God or through the Angel Gabriel. 4th—Belief in God's prophets, the most preeminent of whom were Adam, Noah, Abraham, Moses, Jesus and Mohammed. 5th—Belief in the resurrection and final judgment when all mankind shall appear before God, who will reward or punish them according to the deeds they have done on earth. Opinions differ, of course, as to the nature of these rewards and punishments. 6th—Belief in pre-destination, or the inability of man to avoid, by any act of his own, the destiny irrevocably predetermined by God and written down in the eternal book previous to the creation of the world. At the first glance this seems to deprive man of his character as a free agent, but a closer examination shows that it does not do so. It will be seen, therefore, that in its fundamentals it closely resembles esoteric, or so-called Christianity, when the latter is stripped of its objectionable dogmas. From these central points diverge numerous lines which form in the aggregate an elaborate system of faith and worship, the effects of which vary greatly according to the character of its followers. The articles of religious practice are five in number, namely, Prayer, Ablution, Alms, Fasting and Pilgrimage. And now let us endeavor to ascertain from whence this system sprang, by considering who and what our Prophet was. And let me assure you that in seeking for the truths I have found, I have had to overturn a vast deal of rubbish in the shape of false history, false opinions, and false reasoning, before I caught the faint gleam of that priceless jewel which has been preserved to man through all the ages, although the bigots and pharisees of orthodoxy have striven most earnestly to destroy it. In the light of reason and human testimony it has been quite conclusively shown that Mohammed was a pure and holy man who voluntarily gave up all that the world holds dear, in order to obtain a knowledge of the one great spiritual truth; that he suffered ridicule, obloquy, and persecution of the cruelest, most vindictive character in trying to teach this truth to the Arabs, and that he died in extreme squalor and poverty after having fully accomplished his mission. These are facts which are generally admitted even by Christian writers, and it is, therefore, unnecessary to call Mohammedan witnesses. It is said that a young man once asked Jesus what he should do in order to inherit everlasting life. The reply was: "Sell all that thou hast, give to the poor, take up the cross and follow me."[329]

---

[329] Complete quote: "One thing thou lackest: go thy way, sell whatsoever thou hast, and give to the poor, and thou shat have treasure in heaven:

Now, that is exactly what Mohammed did, except that he did not follow Jesus in the sense understood by the orthodox Christian. He sacrificed all he had in the world and bore the cross of trials and suffering faithfully and patiently until he had securely planted the true faith in the East. It has been plainly shown by every writer who has given us anything purporting to be a record of the life of our Prophet, that during his childhood he was remarkable for his quiet, gentle demeanor, his quick intelligence, and his modest, retiring, thoughtful disposition. While he mingled freely with the boys of Mecca he did not acquire any of their coarse and vicious habits. As a youth he was distinguished by his affectionate ways, his frankness and truthfulness under all circumstances, and as a man he was upright, just and generous in his dealings, and was a merchant whose honesty and reliability were unimpeached. So generally was his character recognized that he was called by the people of Mecca: "Al Amin, the Trusty." Does it seem possible that a man with such pronounced and well-grounded principles of morality and integrity followed until the age of fifty, could suddenly change and become what many of the Christian writers have wickedly declared him to be? I don't believe it. All the prominent Christian authors have been forced to the confession, more or less plainly made, that after protracted study and research, they were utterly unable to form an adequate and satisfying conception of his character. The explanation of their failure is plain—they reasoned from the standpoints of their own materialistic creeds, and facts which would have solved the riddle for them, had they been able to break away from their false opinions and beliefs, were cast aside as valueless. Several of these sapient writers have said in substance: "Mohammed was not our kind of a Christian, and, therefore, he must have been an impostor. But it troubles us to find that such a truly pure and holy man was not a Christian." Had they understood the teachings of their own Prophet they would not have wondered at this seeming phenomenon. It has been admitted that Mohammed's material prospects were all that the most ambitious young man of his

---

and come, take up the cross, and follow me." St. Mark 10:21(KJV)

time could have desired, until he began to teach spiritual truth. His relatives were wealthy, and his uncle, Abu Taleb, who took him into his family and became a kind, affectionate and indulgent father to him after the death of his own parents, was one of the wealthiest and most prosperous merchants in Arabia. The guardianship of the Kaaba at Mecca, the incumbent of which office was also Governor of the city, was hereditarily fixed in his family line, and, had he been content to drift with the tide of life as he found it, would certainly have fallen to him together with his uncle's great wealth.[330] Had he been the crafty, ambitious, unprincipled man he is popularly believed by the Christians to have been, he would undoubtedly have waited patiently and kept in favor with his relatives until, in the natural course of events, he must have become one of the foremost men in Arabia, crowned with wealth and high honors, and surrounded by all the comforts, luxuries and grandeur of a worldly life. But he chose the better way, although it led him through a path thickly strewn with sharp stones and cruel thorns, and, from a worldly point of view, filled his life with bitter disappointments, privations, sorrows and sufferings, of the most harrowing description. And herein is a lesson which should sink deeply into the hearts of those who have turned their backs upon the higher way, and are engaged in that mad chase after rupees and comfort which is so deeply engrossing the attention of the greater part of the world. You are probably all familiar with the history of our Prophet's separation from the affairs of the world. He passed long periods in prayer, fasting and meditation, and reduced his diet to dates and barley, his only drink being water. This abstemiousness was continued to the close of his life, and it is said that sometimes for a whole month he would eat nothing but dates, and those in very small quantities. His favorite place of retirement was a cave on Mount Hara, where he would spend several days at a time in meditation, and where be received his mission to

---

[330] The Kaaba is the holiest structure in Islam and the foci of Muslim prayer (directionally) and pilgrimage. Muslim tradition holds that the stone structure was originally built by Abraham as the first house of worship.

215

shed the light of truth upon the world—to kindle that fire which in after years burned so brilliantly and filled the entire East with its glorious effulgence.[331] He was often accompanied by his faithful wife, who was the first convert to his doctrines, and who seemed to share with her whole soul his zeal and devotion to his purpose. Whenever he emerged from his retirement and returned to his home in Mecca, he went about doing the good that came to his hands and assisting those who, through sickness or misfortune, were unable to provide for themselves. In this way the greater portion of his own wealth, and that which he had acquired through his marriage with Khadijah melted away.[332]

Now, it is necessary to weigh carefully all these well-known and universally admitted facts concerning our Prophet's life prior, and immediately subsequent to, his inspiration, in order to arrive at a rational conclusion regarding his character, and to make those comparisons between him and the other holy prophets which are inevitable. During the period we have now reached, he made no effort to teach publicly any of the truths which had been revealed to him, and his manner of life seems to have attracted very little attention beyond the circle of his immediate relatives. At this time he was looked upon as a harmless fanatic, who had foolishly sacrificed his brilliant worldly prospects for some reason which was not quite clear to his friends, and in which they took no very great interest. Subsequently he publicly announced his mission, and called down upon himself that bitter, relentless, cruel persecution and storm of ridicule which has been more or less graphically described by the historians. Has there ever been a prophet who attempted to teach the world the one true way to eternal life whose path has been strewn with roses? Not one. The world hates the truth with fiendish malignancy, and calls madly for the blood of him who attempts to teach it. The claim made by Mohammed that aroused the fury and contempt of the Meccans was substan-

---

[331] Usually spelled Mount Hira.

[332] Khadijah was the Prophet Muhammad's first wife, who had gained wealth through owning property and engaging in trade prior to their marriage.

tially the same as that made by Jesus of Nazareth, who received similar treatment at the hands of the enraged Jews. He said that he was a prophet and apostle of God, inspired by the Supreme Being to point out to the Arabs the true way to salvation and to redeem them from idolatry and the vices which they had acquired indirectly from following a thoroughly materialistic religion. He repeatedly told his hearers that he was not a supernatural being but a man, like unto them, with the same physical structure, the same mental endowments, and the same natural inclinations and proclivities; that from the Supreme Spirit he had learned the solution of the mystery of life and death and the true way to eternal life; that he had given up the world that he might teach them this way—had crucified himself that they might live—and this is the only way in which Jesus was ever crucified, all the sophistical arguments and quotations from false history to the contrary not-withstanding. Mohammed claimed to be a prophet sent from God in the same sense that Moses, Abraham, Elias, Jesus, and every other truly inspired prophet claimed to have been sent. He taught no new religious system, but sought to revive that one eternal truth which has been preserved to man from the beginning of the world, and will continue to be preserved as long as the world shall stand. His claim was no more nor less than that made by Jesus, who never claimed to be God nor the son of God in the sense in which some misguided people believe he did. In the 58th verse of the 8th chapter of the Gospel of St. John is a statement attributed to Jesus which has sorely puzzled Bible readers and commentators, but which is a plain, unequivocal statement of all that Jesus claimed to be, when it is properly translated from the original Greek. It reads thus: "Jesus said unto them, Verily, verily I say unto you, before Abraham was I am." As the verse stands, it is an ungrammatical absurdity and means nothing. The correct translation into English would make it read thus: "Verily, verily I say unto you, I am what Abraham was before me." That is to say, an inspired prophet like Abraham. Jesus admitted that there were other true prophets before him, and some of the Mohammedan doctors insist, and their arguments and evidence are by no means unworthy of consideration, that he plainly prophesied the coming of Mohammed, declaring that the latter would lead his followers into the truth. And I assure you that our Prophet never taught anything that was at all at variance with the true teachings of Jesus; on the contrary, a careful comparison of the true tenets of the faith of Islam with those taught to his disciples by the Prophet of Nazareth cannot fail to show that in their tone and tendency they are identical. Mohammed frequently referred to the Nazarene as "Jesus, the inspired son of Mary," the Holy Prophet sent by God to the Jews, and paid to him the loftiest tributes of love, reverence and esteem; but for the absurd dogmas, misconceptions and superstitions of the system erroneously called Christianity he had nothing but the strongest condemnation. He taught that at certain periods in the development of humanity a prophet arose from among the people to raise them from the degrading materialism of dogmatic creeds, the unhealthy growths of ambition, selfishness and worldliness, and to guide them into the true path from which the desires of the flesh had led them to stray; that this was God's method of holding humanity in its upward course toward spiritual perfection. He declared that he was the last of a long line of prophets, and that he taught nothing different from the teaching of his predecessors, that his purpose was to renew the one supreme truth in the hearts of his fellow Arabs. The validity of this declaration is apparent when one has some knowledge of the philosophy of Islam. "What!" exclaims the Christian, "and does Islam really contain a philosophy?" Yes, my misguided brother, it is a philosophy as well as a religion, and a pure, perfect, holy philosophy, too. Look about you and see the beauties and wonders of nature, the growth and decay of the trees and flowers, the movements of the planets, the changes of the tides and seasons—all the grand manifestations of nature moving on with steady, majestic regularity under the guidance of an unseen power, which is a dense mystery to materialistic science. The unvarying order which pervades the whole system indicates the power of a master hand. We do not find grapes growing upon cocoanut trees nor figs upon thistles, but each fruit appears and matures upon its own tree; and decays and passes into nothing, if not used for human food. We see the flash of

the lightning and hear the roll of the thunder and the shrill shriek of the cyclone, but the average man knows nothing of the irresistible force behind all these manifestations. Science has tried to explain the phenomena, and has failed utterly, although it has discovered that there are certain fixed laws and conditions which govern them and make it possible to foretell their coming. Like Jesus and every other prophet who has taught the true doctrine of salvation, Mohammed found it necessary to present his teachings in two aspects, or rather to divide them into two parts—one for the very few who could comprehend or assimilate the higher truths, and the other for the masses who were so blinded by worldliness and the bonds of the flesh that they could grasp only the materialistic ideas of forms and ceremonies. The Koran and the traditions are full of suggestions of this idea. Jesus said to his disciples: "Unto you it is given to know the mysteries of the kingdom of Heaven, but unto those who are without, all things are done in parables."[333] If Jesus and Mohammed were inspired by God, it is only reasonable to conclude that the mysteries of what we call "Nature" were unfolded to them, and they were able to teach a higher philosophy than that known to the materialist of to-day. And I tell you frankly, that it was through this exalted philosophy that I was brought to Islam. Every Mussulman who reads his Koran and has any conception whatever of the doctrines of his religion freely concedes that Jesus was an inspired prophet; but he also knows that the system known to-day as Christianity and taught from the pulpits of the so-called Christian churches is no more like the system taught by the Nazarene than is the African Voodoo system or the absurd antics and notions of the Salvation Army. When I talk to a Christian about Islam he promptly declares that it is opposed to civilization, endorses polygamy, has a horribly bloody record, and that its followers are fanatical and intolerant; all this, in his estimation, is true, and ought to condemn any religion.
  The average ignorant Christian in America—and I

---

[333] "Unto you it is given to know the mystery of the kingdom of God: but unto them that are without, all these things are done in parables." Matthew 13: 10-13 (KJV)

suspect he is the same in England—fully believes that every Mohammedan has a harem full of wives where he spends the greater portion of his time when he is not prowling about with sword in hand seeking to kill a Christian. A fairly intelligent Christian once told me that every Mohammedan believed that he could not get a really good place in Paradise until he had killed and eaten a Christian.

"But," said a Parsee to me in Rangoon, "you cannot deny that our present advanced civilization is due to Christianity." "Well I do deny it," I said; and then I referred him to the New Testament and to the Sermon on the Mount, and asked him to try to harmonize those principles with what is called Christian civilization. It cannot be done. The cold truth is that this Western civilization has nothing of the true spirit of Christianity in it, but is the legitimate offspring of ambition and selfishness. It is also a well-known fact that the course of Western progress and advancement has always been obstructed by the Christian Church, ever since that Church has had an existence. It has always stood in front of the procession and shouted with uplifted hands: "You mustn't go any further, or you will weaken and degrade me." And then when it has been pushed aside and the irresistible tide has swept past it, it has tried in every way to impede and harass the moving column. And now with marvelous assurance and impudence it says: "See what we have done. Look at our glorious Christian civilization, and then fall down and worship us." The truth is, and every man can ascertain it for himself, that what is called Christian civilization was born in the 8th century among the Moslems of Spain, while the Christian world was plunged in the depths of ignorance and barbarism.

Professor Draper says: "I have to deplore the systematic manner in which the literature of Europe has contrived to put out of sight our scientific obligations to the Mohammedans. Surely they cannot be much longer hidden. Injustice founded on religious rancour and national conceit cannot be perpetuated forever...The Arab has left his intellectual impress on Europe, as, before long, Christendom will have to confess; he has indelibly written it on the heavens, as anyone may see who reads the names of the stars on a common celestial globe."[334]

Everyone who believes that ours is a Christian civilization, born of Christianity, should read Professor Draper's "Intellectual Development of Europe," or any other honest historical work treating of the subject. Stanley Lane-Poole in his "Moors in Spain" shows that Europe acquired her first knowledge of the arts and sciences from the Moslems, and while showing the degraded condition of the rest of Europe, says: "Whatsoever makes a kingdom great and prosperous, whatsoever tends to refinement and civilization, was found in Moslem Spain."[335] I earnestly hope that the Christians will soon learn to be rational and honest, and that we shall hear no more of this senseless twaddle about Christian civilization.

And now let us glance briefly at that great bugbear, polygamy. Almost the first question a Christian asks me is "Do you believe in polygamy?" "Yes," I reply, "under certain conditions." And last week an educated Mussulman said to me: "Surely you do not believe in polygamy—you would not advocate its introduction in America?" He was clearly surprised when I said that I not only believed in polygamy but would advocate its introduction into the American social system as soon as America had become sufficiently moral and refined to adopt it decently and respectably. Let us look at this question rationally. When our Prophet taught on earth, unlimited polygamy was sanctioned by the laws and social customs of Arabia. He modified the practice by allowing his followers to marry only four wives, telling them plainly that they should not marry any more women than they could treat with equal love and justice. He subsequently declared that no man could love two wives equally, and this declaration practically annulled the privilege of marrying four. It is quite clear that his purpose was to purify the existing social conditions and elevate the domestic relations to a higher moral

---

[334] John William Draper, *History of the Intellectual Development of Europe*, 2:42.

[335] Stanley Lane-Poole, *The Story of the Moors in Spain*, preface.

standard. While polygamy is looked upon in the West as vicious and criminal and subversive of morality and justice, in the East it has quite a different aspect, owing to social traditions and customs, and is considered quite in harmony with the most exalted ideas of propriety. Now there are many sides to this question, and to discuss it fully would require more time than is at my disposal this evening. I freely admit the fact that the introduction of polygamy at once into our American social system would certainly prove most pernicious; but when the system and its purposes and tendencies are properly understood, and the beneficent moral influences of Islam have produced the effects which they must inevitably produce among an educated and enlightened people, it can and should be advocated. It is absolutely the only remedy for the curses of prostitution and marital infidelity with which America and Europe are most grievously afflicted, and it will elevate our womanhood to that exalted and admirable position which it is fitted by nature to occupy. While Mohammedan law permits polygamy, it does not present it as an article of faith, and no good Mussulman would think of marrying more than one wife unless the conditions existing in his home imperatively demanded such a course, and he could conform to the spirit of the Islamic requirements. One is allowed to marry four wives and no more, if in so doing he does not violate the laws of the land in which be lives, but he may have one wife or none at all without sacrificing any of his religious rights or privileges. One can be just as good a Mussulman, as a monogamist or a bachelor, as he can as a polygamist, and will stand just as well in the estimation of his neighbors.

Go with me into any large American or European city and see the evidences of that resistless torrent of vice and crime that rushes and seethes through the social fabric; go with me to a fashionable ball, reception or dinner party and see the position in which noble woman, one of the grandest works of God, has been placed by the usages and customs of this 19th century civilization; see the honored wives of wealthy educated Christians and their virtuous daughters, exposing to the view of men whose blood and passions are fired by the fumes of alcohol, personal charms which should be seen only in the privacy and purity of the home; take up the newspapers and see the records of divorces, social scandals, and marital woes that fill us with shame and disgust, and then tell me that these so-called Christian laws and Christian customs are good things. And where is the remedy for all this? In Mohammedan laws and customs—in Islamic principles. Christian laws and customs have been tried for many centuries and have failed utterly. Now give Islam a trial.

And now let us touch briefly upon the bloody record of Mohammedanism and the propaganda of the sword. So far as blood-stains are concerned I hardly believe that Mohammedanism has any reason to blush when she compares her garments with those of Christianity. Have you ever read the history of the Inquisition and Crusades? When the Khalif Omar took Jerusalem in 637 he rode into the city by the side of the Patriarch Sophronius, conversing with him concerning its antiquities.[336] Not a drop of blood was shed. But when the Christian Crusaders entered it the brains of young children were dashed against the walls, infants were thrown over the battlements, every woman that could be seized was violated, men were roasted on fires; some were ripped open to see if they had swallowed gold; the Jews were driven into their synagogues and there burned. About 70,000 people, men, women, and children, were cruelly butchered. And this is the testimony of Christian historians, not Mohammedan. It has been clearly shown that our Prophet, as well as the first Caliph, Abu Beker, repeatedly and emphatically directed the Moslem generals to refrain from killing or injuring women, children, and old men; not to destroy the fields of grain or date trees, and to sheath the sword at once upon the surrender of a city.[337] Humanity, mercy, and kindness were insisted upon.

---

[336] Umar ibn al-Khattab, 581?-644, second caliph of Islam who lead several campaigns in the early expansion of the Islamic empire, including the holy lands. *Encyclopaedia of Islam*, New ed., s.v. "'Umar (I) B. Al-Khattab." Sophronius, 560?-638, Patriarch of Jerusalem (634-638). *New Catholic Encyclopedia*, 2nd ed., s.v. "Sophronius, St."
[337] Abu Bakr, 570?-634.

When the Prophet entered Mecca, upon its surrender to him, not a man, woman, or child was killed or ill-treated, and not a single house was plundered, although this was the city in which he had been so shamefully abused and persecuted, and its inhabitants were those who had cruelly wronged him. Why did he not take revenge when the opportunity offered itself? There was not a single sentiment of revenge or malice in his whole being; he was a Prophet of the God of love, truth, justice, and mercy.

Both records are sad enough and bloody enough, Heaven knows, but I am firmly convinced that there is far less fiendishness and blood-thirstiness, and bestiality to be answered for by the Moslems than by the Christians. Did the Christians have a valid warrant in the teaching or conduct of the meek and lowly Jesus, giving them authority to go about murdering, in cold blood, those who did not believe as they did? But of course, they don't do it any more; not because some of them wouldn't like to, but public opinion has changed. It isn't safe to be too enthusiastic and conscienceless and brutal now in making proselytes to one's religion no matter how confidently one may believe in it himself. And when I say to you that Mohammed never advocated, taught, nor consented to the propagation of Islam by means of the sword, and that he severely condemned violence and taking of life in any form, I tell you truths which can readily be verified by any honest, unprejudiced person who will take the trouble to investigate the matter impartially.

A learned Moslem writer says:—"The remark that the sword is the inevitable penalty for the denial of Islam is one of the gravest of the false charges imputed to this faith by the professors of other religions and arises from the utter ignorance of those who make the accusation. Islam inculcates and demands a hearty and sincere belief in all that it teaches, and that genuine faith which proceeds from a person's heart cannot be obtained by force or violence."[338]

The Holy Koran says:—"Let there be no forcing in religion; the right way has been made clearly distinguishable from the wrong one. If the Lord had pleased, all who are on

---

[338] Quote not identified.

224

the earth would have believed together; and wilt thou force men to be believers?"[339]

Our Prophet himself was as thoroughly unaggressive, non-combative and peace-loving as the typical Shaker, and, while he realized the fact that a policy of perfect non-resistance would speedily have resulted in the murder of himself and every Moslem in Arabia, he urged his followers to avoid, as far as possible, violent collisions with the unbelievers and not to fight unless it was necessary in order to protect their lives. There are a number of accusations made against Mohammedanism, which, even if true, cannot justly be said to have even the remotest relation to the doctrines of Islam; there are zealots and fanatics in all religious bodies, and it is due to their weaknesses that discredit falls upon the faith they profess to follow.

It would be useless to attempt, in a single lecture, to reply to all the false charges made against Mohammedanism by ignorant and prejudiced writers; but at the risk of exhausting your patience, I will refer to one more. It has been said that toleration is unknown among the Moslems. A Christian writer in "Chambers's Encyclopaedia" says:—"One remarkable feature of the Moslem rule in Spain deserves mention, as it contrasts them so favorably with the contemporary and subsequent rulers of that country even to the present time, and that is their universal toleration in religious matters." Being a Christian this writer can hardly be accused of partiality to Islam.

Godfrey Higgins, another writer who can safely be called a 19th century Christian at least, says:—"Nothing is so common as to hear the Christian priests abuse the religion of Mohammed for its bigotry and intolerance. Wonderful assurance and hypocrisy! Who was it that expelled the Moriscoes from Spain because they would not turn Christians? Who was it that murdered the millions in Mexico and Peru and gave them all away as slaves because they were not Christians? What a contrast have the Mohammedans exhibited in Greece! For many centuries the Christians have been

---

[339] Two suras of the Qur'an have been combined. The first sentence is from Sura 2:256 and the second Sura 10:99.

225

permitted to live in the peaceable possession of their properties, their religion, their priests, bishops, patriarchs, and
churches; and the war between the Greeks and Turks was no
more waged on account of religion than was the war between
the negroes in Demarara and the English...Wherever the Caliphs conquered, if the inhabitants turned Mohammedans,
they were instantly on a footing of perfect equality with the
conquerors."

An ingenious and learned dissenter, speaking of the
Saracens says:—"They persecuted nobody; Jews and Christians all lived happy among them."

Higgins also says:—"In all the history of the Caliphs
there cannot be shown anything half so infamous as the Inquisition, nor a single instance of an individual burnt for his
religious opinion; nor, do I believe, put to death in a time of
peace for simply not embracing Islam."[340]

"But," says the Christian, "all that was characteristic
of the centuries long past—Christians are not so bigoted and
intolerant to-day." Aren't they? Go to the Philippine Islands,
a country with a population of over seven millions of souls,
which is and has been under the rule of Christian Spain for
300 years. Try to teach any religious system but the Roman
Catholic and see what will happen to you. There is no Mohammedan country on earth that refuses to admit Christian
missionaries, and that does not give them ample protection.
Three years ago two representatives of the British and Foreign Bible Society of London went to Manila, the capital city
of the Philippine Islands, to sell Bibles. One of them died
within three weeks after his arrival, and there were people
wicked enough to say that he was poisoned at the instigation
of some of the Catholic priests. The other was arrested, and
thrown into jail on a charge of teaching contrary to the State
religion, but was afterwards sent to Singapore by order of the
Spanish Government. That was only about three years ago. A
few months later seven Buddhist priests from Foo Chow,
China, who were induced by their fellow-countrymen in Ma-

---

[340]Godfrey Higgins, 1773-1833, author of *An Apology for the Llife &
Character of the Celebrated Prophet of Arabia, Called Mohamed, or
the Illustrious* (1829). Quotes come from this work.

nila to go there believing that they would be allowed to celebrate a Buddhist ceremony there, were arrested, fined, and sent back to China. A vast volume of evidence might be adduced to show the utter groundlessness of the charge of intoleration. As a matter of fact, intoleration is entirely foreign to the principles of Islam, and no intelligent Moslem would be guilty of, or consent to it.

The essence of the true faith of Islam is resignation to the will of God, and its corner-stone is prayer. It teaches universal fraternity, universal charity, universal love, and universal benevolence, and requires purity of mind, purity of action, purity of speech and perfect physical cleanliness. It is the simplest and most elevating form of religion known to man. It has no paid priesthood, nor elaborate ceremonial, admits no vicarious atonement, nor relieves its followers of any of the responsibility for their sins. It recognizes but one God, the Father of all things, the divine spirit that dwells in all the manifestations of nature, the one omniscient, omnipotent, omnipresent ruler of the universe, to whom its followers devoutly pray and before whom all stand upon a platform of perfect equality and fraternity. The devout Mussulman, one who has arrived at an intelligent comprehension of the true teachings of our Holy Prophet, lives in his religion and makes it the paramount principle of his existence. It is with him in all his goings and comings during the day, and he is never so occupied with his business or worldly affairs that he cannot turn his back upon them, when the stated hour of prayer arrives, and present his soul to God. His loves, his sorrows, his hopes, his fears are all immersed in it; it is his last thought when he lies down to sleep at night and the first to enter his mind at dawn when the voice of the Muezzin rings out loudly and clearly from the minaret of the mosque, waking the soft echoes of the morn with its thrilling, solemn, majestic monotones: "Come to Prayer! Come to Prayer! Prayer is better than sleep! Prayer is better than sleep!"

X.

HYDERABAD LECTURE— PHILOSOPHIC ISLAM

November 25, 1892

It has been suggested to me that, inasmuch as I am to address an audience composed almost entirely of Mohammedans, many of whom are men of unusual mental culture and attainments, I should devote my remarks to the philosophical aspect of Islam rather than to its popular expression in devotional and social laws and forms. I certainly cannot hope to tell anyone here anything which he does not already know of Islam as an esoteric religious system; nor would it be possible for me to teach the Islamic philosophy in a dozen lectures even if I should be bold enough to attempt to teach it at all.

Now I am fully aware of the fact that there are many professed Mussulmans who do not know that there is a philosophic side to their religion; and perhaps it is just as well that they do not, for such knowledge might, possibly, lead them away from the plain, safe and simple truths already within their grasp, and out into the broad and dangerous ocean of metaphysical speculation where their frail mental barks would be wrecked upon the rocks of doubt and despair. But in this age of rapidly increasing scepticism and doubt, the legitimate offspring of advanced materialistic education and civilization, there are many minds which require more evidence of the truth of a religious system than the dogmatic assertions of those who claim to be religious authorities. The simple, child-like faith handed down to them by their ancestors has been blighted by the cruel frosts of atheism and agnosticism or frozen to death by our Western nineteenth century civilization.

Now in some respects this condition is a direct and positive benefit, for when the congealed plant thaws out under the warm sunshine of spiritual truth, it is very liable to grow and fructify in a manner impossible to it under the old influences. I will not say that it would have been better if the works of many of our modern materialistic philosophers had never seen the light of day, for it is quite probable that they may prove, after all, very potent factors in bringing the educated, progressive masses into the true path; but it does seem at times as if they were tearing down where there is little or no hope of rebuilding. I am heartily in sympathy with the man who demands a reason for everything, for that has been my condition of mind for twenty years. I abandoned the system improperly called Christianity, soon after I attained my majority, because its teachers could not give me a convincing reason for the faith that was in them.   When   I   was asked to believe that one was really three, and that a just and merciful Creator committed an act of unnatural cruelty simply to gratify a mere whim, I demanded corroborative testimony and was told that none had ever been filed. This very grave omission compelled me to throw the case out of court.

But it seems to me that every rational man should make an earnest, honest, unprejudiced effort to get at the bottom facts concerning his religion, and ascertain to his own complete satisfaction whether it is a mere jumble of whimsical, purposeless forms and ceremonies, or whether it is based upon the eternal truth. If the atheist and materialist are correct in their conclusions, then there is no reason in the world why man should have a religion at all; but it seems to me that the question is of sufficient importance to man to justify more than a superficial examination. If, as is claimed by the religious world, this present life is but an infinitesimal period as compared to the life beyond, and that the conditions of the future existence are largely determined by our course of conduct and thought here, then it is of vital importance to us to know whether the claim of religion is valid and what kind of a religion is necessary in order to bring about the desired results.

And what is religion? Cicero defined it as "that which brings with it the care and cult of some higher power which men call divine."[341]

Max Müller, in his Gifford lectures, after carefully tracing the history of the word, finds that its earliest conception can have meant only respect, care, and reverence; that later on it took the moral sense of scruple and conscience and, lastly, became more and more exclusively applied to the inward feeling of reverence for the gods and to the outward manifestation of that reverence in worship and sacrifice.[342] What the Romans expressed by "religion" was chiefly the moral or practical, not the speculative or philosophical side of religion. The Greeks at first used the word to express fear of the gods or demons and the divine power in a good sense. But very soon it began to be used in a bad sense as expressive of superstition. Marcus Aurelius spoke of it as meaning God-fearing without superstition.[343] Spinoza thinks that practical religion ought always to be simple piety and obedience as distinguished from philosophy and love of knowledge.[344]

The modern definitions are numerous and varied. One philosophical writer says: "Religion is our recognition of the unity of nature and teaches us to consider ourselves as part of the whole."[345]

---

[341] Marcus Tullius Cicero, 106-43 B.C. Quote from Cicero, *De Inventione*, Book II, Ch. 53.

[342] Max Müller, 1823-1900, delivered his lectures on "Natural Religion" before the University of Glasgow in 1888. Lectures two through five dealt with the various definitions of religion. F. Max Müller, *Natural Religion*.

[343] Marcus Aurelius, 121-180, Roman Emperor, Webb refers to ideas from the work *Meditations*.

[344] Baruch Spinoza, 1632-1677, Webb paraphrases Spinoza's ideas from *Tractatus Theologico-Politicus* (1670).

[345] Quote not identified.

Another writes: "Theology and metaphysics have nothing to do with morality. Religion has never been other than science plus worship or devotion."[346]

Max Müller says: "We can hardly open a book without meeting with random definitions of religion. Religion is said to be knowledge, and it is said to be ignorance; religion is said to be freedom, and it is said to be dependence; religion is said to be desire, and it is said to be freedom from all desires; religion is said to be silent contemplation, and it is said to be splendid and stately worship of God. People take every kind of liberty with this old word. Young poets will tell you that poetry is their religion; young artists that their religion is art; while it has been said of old that pure religion is to visit the fatherless and widows in their affliction and to keep yourselves unspotted from the world."[347]

In the consideration of the subject we will find it convenient to keep in sight the dividing line between Religion and Theology, words which are too often used promiscuously. By "religion" we should understand the subject; by "theology" the study or science of that subject.

"Religious ideas of one kind or another," says Herbert Spencer, "are almost universal. We are obliged to admit that if not supernaturally derived, as the majority contend, they must be derived from human experience slowly accumulated and organized. Considering all faculties to result from accumulated modifications, caused by the intercourse of the organism with its environment, we are obliged to admit that there exists in the environment certain phenomena or conditions which have determined the growth of the feeling in question, and so are obliged to admit that it is as normal as any other faculty. We are also forced to infer that this feeling is, in some way, conducive to human welfare."[348]

Now this frank admission affords us a clue to a bit of evidence which will prove of value to us in our investiga-

---

[346] Quote not identified.

[347] See Müller, *Natural Religion*, 43-44.

[348] Herbert Spencer, *First Principles*, Part I, Ch. I, Sec. 4. A mix of quotes.

tions, for it will naturally lead us to consider the religious instinct in man and its probable and possible causes. In every age, every clan, tribe and nation has had its idea of religion and its distinctive form of worship; and it would be extremely difficult, not to say impossible to-day, to find, outside of civilized Christendom any collection of men without some sort of form for the expression of their ideas of religion. There are in some of the islands of the Pacific Ocean, tribes which have never, so far as is known, come in contact, or had direct and intelligent intercourse with people professing any of the popular forms of religion; and yet they have religious ceremonies and ideas of post-mortem existence which are near enough to those to be found in the Mohammedan, Christian and Buddhist systems to warrant the suspicion that they might have been borrowed from those systems. There are, in the mountainous regions of the Philippine Islands, savage tribes who have never been subdued by the Spaniards, who have the most intense hatred for a white face, and who know of no other country or religion than their own, but who still have religious forms and religious ideas very similar to those of some of the earlier and more superstitious Christian sects.

What is the cause of this religious faculty, which the evidence at hand proves to be just as normal as any other faculty? It cannot be the result of mere chance nor of education, for it can be clearly traced back through all generations of mankind as far as history leads us, and it is as clearly shown that it was strongest where education was unknown. It is just as much a part of human nature in its uncultured, unrefined condition as it is in its most refined aspect. In fact, we find that there is, as a rule, less religious devotion among the educated than among the uneducated classes.

And now, having reached this point in our journey toward the ideas which I desire to bring to your careful attention, let us turn aside into a familiar path and see whether materialistic science can answer this question and some others which we may find it necessary to ask her. We often hear of the irrepressible conflict between science and religion, but that can only mean the conflict between materialistic religion and materialistic science, not between true religion and true

232

science. All efforts to make materialistic creeds harmonize with materialistic science always have been, and always will be, utterly futile.

Without stopping to analyze the popular creeds or to consider what science knows, let us see what the latter does not know. We see the trees growing and the flowers blooming under a steady, unvarying power that scientists call an impenetrable mystery. They have never yet been able to tell us what life really is, whether or not the soul really exists and, if it does, what becomes of it upon the death of the body. The educated physician can tell you the name of every bone, muscle, nerve and organ of the human body, and the location and probable function of all, except the spleen, and yet he is densely ignorant of the power that makes the body a living, breathing man, full of love and hate, and all the passions and inclinations of humanity. A thrust of a knife or a bullet in the heart or brain converts that animated form into an inert, lifeless mass which speedily becomes a heap of festering corruption breeding new life in the form of worms and insects. Can the scientist make the man live and love and hate as he did before? No; some thing has gone out of him never to return. What is that something, and where has it gone? Science is dumb. Can science tell us what electricity is? No. And yet in the Western world, particularly in America, we have gathered this subtle, mysterious fluid and made it serve us in various ways. We have harnessed it to our street cars, and have forced it to shed its light in our streets and homes. We have even made it carry the human voice for hundreds of miles, and written messages for thousands. And what is it? Science does not know. Science has learned to foretell with reasonable accuracy the coming of the storm, but it cannot tell us what the force is that produces the roaring, shrieking cyclone nor the gentle wind that fans our brows. It does not know what sleep is, nor why we dream. It cannot even tell us how and why we think. Science has done much for this materialistic civilization of ours, and it moves about with a confident, lordly air among those things which it can cut, and melt, and test chemically; but when it stands in the presence of the wonders of life and death and views the operations of those wonderful laws which govern the various manifestations of what we call "nature" it drops its hands and head in helpless confusion.

And yet our sapient philosophers continue to talk seriously and earnestly about the conflict between science and religion, just as if science had really learned all there is to be known about man and nature. Compare what science knows with what she does not know, and you will see that she is altogether too ignorant to serve as a guide to anyone who desires to know the higher truths. And yet let me assure you here that there is no known fact in modern science that is not strictly in harmony with the principles of that higher science which is known to but few in these degenerate days.

Do you know what thoughts are? Did you ever try to analyze the motives that prompt the various acts of your lives? A very wise philosopher says:

"It is impossible to define motives accurately, even our own. We cannot say sometimes why we do a thing. Every reason may be against it—common-sense, habit, inclination, experience, duty—all may be pulling one way, and yet we tear ourselves loose and do the thing."

Is this not true? Sit quietly and watch your thoughts as they come unbidden and of their own accord before you, like so many phantoms. Try to catch one and hold it in your mind steadily for ten minutes—or even five—if you can. Before you have held it a minute another will crowd it out of your grasp and slip away from you before you can recover yourself. Are you master of your thoughts? Can you control yourself?

But I fancy I hear someone say: "What is the use in talking about these things, when they are mysteries which no man can solve?"

They are not mysteries which no one can solve. It is true that materialistic science cannot solve them, but that is because she is working on lines which really carry her away from the truth, and steadily and persistently avoids the methods that have been clearly pointed out to man. Our Holy Prophet—peace be with him forever—knew all the mysteries of life and death, and so did Jesus of Nazareth, and so did every other truly inspired prophet the world has ever known. The one truth—the only real science—has been handed down from the beginning of human development, through the long line of prophets to Mohammed—*Hazrat Moham-med Sallal laho aliehi wasullam*[349]—and this truth has been offered to mankind earnestly and eloquently, while the masses have turned their backs upon it and have wandered off in search of money and comfort and pleasure.

Let us pause and consider one important fact. The teachings of the prophets, as well as those of materialistic science and our own daily experience, lead us to the conviction that we are different from the animal world. Man, with his freedom of thought and action, his intellectual faculties, his fine discriminating powers, is, and should be, above the animals although he has the animal instincts within him to a greater or less extent. But is he so very much different from the animals in his general motives and disposition? Let us see. The spider weaves its web in a convenient place where some unlucky fly may be caught in it. What is the spider's motive, so far as we are able to conceive of it? Do you believe that he stops and calmly and deliberately reasons that the body of the fly will assist his own physical development if he eats it? The most logical and reasonable inference is that he doesn't think about the matter at all, but is forced by an irresistible impulse to catch the fly and eat it, or that he has discovered that flies taste good. From his general conduct we believe that he is incapable of reasoning. Suppose that all the spiders in the world should suddenly become possessed of the impulse to refrain from eating—the development of the species would cease and it would become extinct. Now, the human spider is different—he is capable of reasoning; and when he weaves his web in which to catch a human victim he calculates that the victim will yield so many rupees. What does he want the rupees for—to do good with, to make his brother-man happier and better? Sometimes; but usually be wants the rupees to buy comfort and pleasure for himself, and to enable him to gratify his own animal tastes and instincts. The cow and horse are impelled to eat and drink, not as the result of any regular, consecutive

---

[349] Translates roughly from Urdu and Arabic as "His highness Mohammed, Allah's peace and blessings be upon him."

reasoning, but by that irresistible impulse which compels the animal world to push forward its physical development. Man, on the contrary, beside the instinct of physical development, has also the intellectual and higher spiritual faculties. Is it his first aim and purpose in this life to develop these faculties, or does he yield himself willingly to the animal instinct and make his life's motive the acquirement of those things which he believes will give him bodily comfort and pleasure? If you will ask the young man who is striving to secure a perfect and complete education why he is doing so, he may tell you that his first purpose is to fit himself to be of service to humanity; but isn't it quite probable that his real motive is to win the respect and admiration of the world, to secure a good social position and just a little money in the future? Let me assure you here that, I am not trying to depreciate education and the development of the intellectual powers. I want to try to induce you to analyze yourselves and to learn the general drift of humanity. Our Prophet once said that he who knew himself knew God; if you will study yourselves you will certainly find the subject interesting, to say the least.

"And what has all this to do with our religion and the mysteries of life and death? " someone may ask.

Everything. It brings you face to face with the fact that science, modern science, which the average educated man regards with awe and admiration, and which many consider not only reliable, but infallible, is utterly powerless to explain the most important problems of human life, and is forced to confess its inability to do so. It also offers suggestions calculated to lead the thinking man to believe that there may be ways of acquiring this valuable knowledge that science has never heard of.

Is it worth any man's time and effort to ascertain whether there is really any conscious life beyond the grave, and what it is necessary for him to do in order to secure the best condition that life can afford? That is a question which everyman must answer for himself. If it is true that there is a future life, the inference is only reasonable that it must cover a very much longer period than the sixty or seventy years we have upon this earth. Hence, it seems to me the most important subject that can possibly engage the attention of anyone.

"But," says the sceptic, "of course we want to find out all these things; but how are we going to do so? We have read John Stuart Mill, and Huxley, and Tyndall, and Darwin, and Fichte, and everything else that we could find, and still we are at sea.[350] We have appealed to science, and she not only admits her ignorance and says that no one can know the mysteries of life, but she actually proves, by her own methods for which, by the way, we have great respect, that our religion is not in harmony with her conclusion. What are we to do?"

Now what would you think of a man who tried to take a screw out of a piece of hard wood with a gimlet and insisted upon it that the thing couldn't be done because the tool wouldn't fit into the slot of the screw? Wouldn't you quite naturally and properly call him an idiot and tell him to go and get a screw-driver?

Now modern science has found that its gimlet will not turn the screw of spiritual dynamics, and therefore declares that it cannot be turned. Now what rather surprises me is that the sceptic clings so tenaciously to that gimlet when there is a screw-driver so near at hand.

This is rather a homely comparison, I admit, but I think it expresses the situation better than anything else I can think of at present.

Every true prophet the world has ever known has pointed out the way to the solution of the problems of life and death and every sceptic and materialist has stubbornly and foolishly persisted in ignoring it and following the path which leads directly away from it. He clings to the gimlet and refuses to take the screw-driver when it is offered to him. And why? Simply and solely because if he follows the true way he must give up some of the pleasures and comforts of this earthly life—and, above all, he is afraid that his

---

[350] John Tyndall, 1820-1893; Charles Darwin, 1809-1882. Others mentioned above.

friends and neighbors will laugh at him and avoid his society. Is this not true?

Suppose that some bright, intelligent, popular young man should say to his friends: "I am going to be perfectly pure, perfectly unselfish, perfectly good and perfectly holy; I am going to attend carefully to my religious duties and worship God with all my heart and soul; I am going to tear myself away from the animal side of my nature and cultivate and develop only the higher attributes which God has given me; I will love my fellow-man as if he were my brother, and do my whole duty to God and man."

How long do you believe that that young man would be popular with his friends and associates? Do you believe that they would seek or even tolerate his society very long after he had adopted this course of life? The truth is that very few people are willing to make this sacrifice even to learn that higher knowledge which they would all like to acquire. When you strike at a man's comfort, pleasure, ambition or pocket, he will either hastily beat a retreat, carrying his treasures with him; or fight like a demon to retain them.

There comes to my mind the picture of a grand and noble figure in history—one of the grandest and noblest that man has ever known; a calm, majestic man, in whose personality there is something that commands respect, admiration and awe. His countenance beams with the glory of divine inspiration as he stands with his back against one of the rough pillars of the rude little mosque in Medina, in the erection of which his own hands assisted. Around him are seated a company of men who listen with eager attention to the words that fall from his lips, while they gaze upon him with rapt reverence and devotion. He has given up all that men of the world prize and toil for, has submitted to the cruelest, most inhuman insults and persecution that could be invented by wicked, selfish people who were formerly his admirers and friends; has endured pains and sorrows and disappointments that would have utterly broken an ordinary man. And yet in his heart there is no malice, no desire for revenge, no selfish ambition, no hatred; his soul is full of peace and joy and love, for the divine light pervades his whole being. Patiently and earnestly he points out to his humble followers the true way to eternal life, while they listen gratefully and attentively, letting his words sink deep into their hearts, and treasuring them there in all their daily lives. They do not question the truth of his teaching—they do not doubt—but say: "Show us the way and we will walk in it trustfully and loyally until we have found the priceless jewel of eternal truth."

And what is the way pointed out by this grandest of all prophets? Islam! Resignation to the will of God; the omniscient, omnipresent, omnipotent God, who stands ready to lead the aspiring soul out of the darkness of materialism into that light which shines for all as a guide to Paradise. The way has been clearly pointed out, and if man will not follow it he cannot expect to see beyond the limits of the material world.

The Holy Koran tells us that God is nearer to every man than his jugular vein;[351] that "He guideth not unbelieving people;"[352] that "He guideth whomsoever He listeth."[353] What does this mean when taken with that other saying of our Prophet: "He who knows himself knows God?"[354] The meaning must be clear to every intelligent person.

Materialistic science has discovered, as many a man who knows nothing about science has learned from his own experience, that the continued exercise of any organ of the human body will strengthen that organ, and that protracted rest will weaken it and ultimately render it useless. If I rest my right arm on any object long enough, or allow it to hang quietly at my side, it will become atrophied and I will be unable to raise it at all. The same law governs the intellectual faculties; if I use my memory and by practice obtain perfect control over it, it will become phenomenally strong in time. How do we learn languages? It is a well-established fact that it is much more difficult for a man to learn a language after he is 40 years of age than it is when he is younger. What is

---

[351] Qur'an, 50:16.

[352] Qur'an, 2:264.

[353] Qur'an, 2:142.

[354] This quote is often used in the Sufi traditions, but is considered by most collections of Hadith (traditions of the Prophet) as not established as authentic.

the reason? Because in his younger years he is exercising those parts of his physical and mental organism that are destined to control his character, and is acquiring powers and forces that it is exceedingly difficult to overcome. During youth these powers and forces are forming, and knowledge of languages readily joins with other kinds of knowledge. We very soon acquire habits good or bad, and often they become so strong that they pass beyond our control. There are many people who are absolutely slaves to liquor and tobacco—they cannot refrain from using them, and seem to be utterly helpless victims to these habits. When a young man first begins the use of whiskey or tobacco it is quite an easy matter for him to break away from them, but after he has used them for years it is quite another matter. I am aware that this condition is considered by some physicians as wholly physical, but I am firmly convinced that it is largely mental, as most of the ailments that afflict humanity usually are. Let anyone who has, tried to give up a bad habit study his own sensations, and see what the result will be.

The human body has no sensation after the soul has left it; it is quite clear that sensation does not originate in it, but is manifested only while it is occupied by the soul. Put a live coal upon the flesh of a living man and he will wince with pain, but you may cover the corpse with coals and there will not be the slightest evidence of sensation. It is the soul that thinks and feels, not the body. The body of itself without the soul cannot crave liquor nor tobacco. Does the craving, cease when the body dies? No, not if the soul is immortal, and, personally, I am fully convinced that it is. Now, suppose we accept the hypothesis that this craving clings to the soul after death, and it has no body through which, or by means of which, it can gratify or satisfy it; would not the suffering be intensified a hundred-fold? Now, this is only a hypothetical case and should not be taken in any sense as a declaration of truth; I advance it simply as a subject for speculative thought. My purpose has been to show by a chain of logical reasoning that the hidden elements of our being are capable of development and cultivation quite as readily and fully as any of the parts of our physical organism. If this be true, one can obtain an idea of the results which may follow the direction of the mental force away from the lower or animal parts of our nature and toward its higher or spiritual parts. The cultivation of fraternal love, perfect cleanliness of mind and body, and devotion to God, must certainly develop those higher faculties of our nature to a degree which will enable them to dominate our whole being.

I think I have said enough to show that it is not the blind, slavish following of the forms and practices of religion that is effective, but that the purpose, the motive, must come from the heart, and that in order to reach the sublime heights of knowledge, one must enter into the spirit of his faith with zeal and earnestness. If man will not yield himself to the guidance of God, but turns himself against the higher spirit within him, and follows the lower elements of his nature, his development will be physical and downward, instead of spiritual and upward. He will be of those of whom the Holy Koran says:

"God hath put a seal on their hearts and their hearing, and over their eyes is a veil, and for them is grievous punishment."[355]

"Their case is like unto the case of one who kindleth a fire and when it hath lighted up that which is around him, God taketh away their light and leaveth them in darkness, so they do not see."[356]

I have repeatedly been asked the question: "How do you propose to introduce Islam into America?" As if it was a task in which I alone was interested and in which God had no part. Some people have expressed doubt as to the ultimate success of the Islamic mission in my country, saying that the people were too progressive and too highly developed intellectually to receive and follow it. This doubt is born of ignorance of the true principles of Islam. One of my reasons for believing that the Americans will accept Islam, when it is properly and fairly presented to them is, that they are progressive and intelligent. The idea that Islam is a system adapted only to the minds of the ignorant, superstitious and partially developed portions of humanity is certainly most

---

[355] Qur'an, 2:7.
[356] Qur'an, 2:17.

erroneous. It is the one eternal truth and not only commends itself to the reason of man, but satisfies the longings of his soul for higher and better conditions.

"But," said a doubter to me in Bombay, "do you expect that the Americans will pray five times a day and fast and adopt polygamy and the purdah system, and do everything else that the Mohammedans of the East do?"

I know that there are many thousands of well-balanced people in America who are capable of understanding and appreciating what is good and reasonable, and I believe that the power and influence of Almighty God may be felt there just as strongly as it can in India or any other part of the world. We propose to present Islam to the people as plainly and logically as it is possible to present it, and to leave the results of its practical operation with God. I know very little of the practical application of the purdah system or of polygamy in the East, and, therefore, cannot say whether they are practiced in the true Islamic spirit or not. If they are not applied properly and justly they cannot produce good results; but in my humble opinion the purdah system and polygamy, rationally and intelligently engrafted upon our social system, are the only possible remedies for the evils with which it is afflicted to-day. Prostitution, marital infidelity, drunkenness and kindred vices are prevalent from one end of the vast country to the other. Orthodox Christianity and orthodox Christian laws have fought in vain against these evils for a hundred years, and still they have steadily increased. Now, I believe that Islam and Islamic laws should have an opportunity to try and rid our social system of the monsters of sin that are preying upon it.

I honestly believe that within five years we will have a Moslem brotherhood in America very strong numerically and composed of just as earnest and faithful Mussulmans as the world has ever seen.

For the past ten years I have carefully watched the course of religious thought in my country and have been in a position which enabled me to view the field to advantage. I have seen the masses of intelligent people drifting away from the Christian churches and forming themselves into free-thought societies, ethical culture societies, non-sectarian societies and numerous other organizations the purpose of which is to seek religious truth. Beside these, there are the Spiritualists, the Theosophists, and an infinite number of other smaller bodies which follow no religious system.[357] Then, too, there are the Unitarians, who, I am satisfied, will adopt Islam when they really know what it is.

I believe that the strongest reason why Islam is not the prevailing religious system in America to-day is because it has been so grossly misunderstood and misrepresented by those Christian writers who have attempted to present it to the world in the English language. The masses of the English speaking world know nothing at all of Islam, except what they have acquired from such prejudiced Christian writers as Sale and Irving.[358] Ask almost any American Christian if he knows who and what Mohammed was and what the Islamic system is, and he will promptly answer yes. But when you come to question him as to the sources of his information he will tell you that all he has read upon the subject is Sale or Irving, or both, and the letters that some of the misguided Christian missionaries have sent home from the East. The most absurd and impossible tales about the Mohammedans in circulation among our Christians have emanated from the missionaries who seem to see all the faults of Mussulmans around them and none of their virtues. If a Mohammedan commits a crime or expresses an absurd superstition, they send a report of it home labeled: "An Article of Mohammedan Faith."

Some years ago there was a fanatical Christian in America who thought that God had called upon him to sacrifice his youngest daughter to prove his faith. He tied the child on an altar after the manner in which be supposed that Abraham tied his son, and plunged a knife into its heart. He afterward said that he fully expected God to interfere in time to save the victim's life as He did in Abraham's case. Of course he was arrested, tried for murder and sent to a lunatic asylum. I never heard that the report was circulated among

---

[357] For further discussion , see Webb, "Islam and Theosophy."

[358] George Sale, 1697-1736, translated the Qur'an in 1734; Washington Irving, 1783-1859, wrote *Life of Mahomet* (1849).

243

Mohammedans that Christians believed in killing their children as sacrifices to God. But let some unbalanced Mohammedan commit an insane act and the chances are that the Western Christians will be told that such acts form a part of the Mohammedan faith.

Now it is errors of this sort, as well as many others that have crept into the Western mind, that we shall seek to over turn. Our plan of operation includes the establishment of a weekly high-class journal for the explanation of the Islamic doctrines as well as the dissemination of general information relative to Mohammedans and Mohammedan social laws; a free library and reading-room, a book and pamphlet publishing house, and a lecture-room where lectures will be delivered once or twice a week, or as often as the circumstances seem to warrant.

In carrying out this project I want to feel that the prayers and good wishes of the Mussulmans of the East are with us; that they are interested in the work and will do all they can to help it along. I want to take your hand and carry it across the sea to be seized in an earnest, fraternal grasp by the people of America. I want the Mussulmans of the East to be united with those of the West in that true spirit of fraternity which our Prophet so plainly taught. The Moslem brotherhood should extend to the four corners of the earth, not in name only, but in the true spirit of perfect unity.

There was a time when Islam was the glory of the world, when it was the centre and inspiration of all that was grand and noble and exalting; the heart and soul of the arts and sciences; the civilizer and teacher of mankind. Will history ever repeat itself in this respect? I hope so.

XI.

MADRAS LECTURE—THE BETTER WAY

December 11, 1892

A few months ago I was talking with an Englishman regarding the relative merits of Mohammedanism and the system erroneously called Christianity, as effective methods of securing salvation. He had firmly declared his belief in a conscious existence after the death of the body, and also averred that Christianity was the best and only system known, through and by which man could attain to that post mortem condition known to Christians as the Kingdom of Heaven. I imagined that he was not really satisfied with the Christian system, and I endeavored to show him some of its defects which are plainly apparent when the light of reason is turned upon it. After I had finished he said:

"There is a great deal of truth in what you say of Christianity; but can you show me a better system?"

Now, that is a question which has been asked by thousands upon thousands of thoughtful, honest, broad-minded people who follow Christianity in a half-hearted, doubtful way, simply because they know of no other system, or, rather, because their prejudices and habits of thought have prevented them from examining carefully the character and tendencies of the one true religion—Islam. There is no religious system known to humanity that is, and has been for centuries, so thoroughly misrepresented and misunderstood, by so-called Christians, as that taught by our Prophet. The prejudice against it is so strong among the English-speaking people of the globe, that even the suggestion that it may be true and, at least, worthy of a careful, unprejudiced investigation is usually received with a contemptuous smile, as if such a thing was too palpably absurd to be considered seriously. It may surprise you to hear that it is generally and quite firmly believed among the Christians of America and Europe that Mohammedanism has been clearly and satisfactorily shown to be a false system, followed by blood-thirsty fanatics, and that all its tendencies are materialistic and degrading. It would astonish and amuse you to know of some of the absurd ideas that have taken hold of the Anglo-Saxon mind regarding Islam and its followers. And yet you cannot justly blame us Americans nor our European cousins for having these ideas, for they are the results of centuries of malignant misrepresentation born of spiteful jealousy, gross ignorance and wicked fanaticism. The early fathers of the Christian Church lied most vigorously and persistently about our Prophet and his teachings, as well as about the Moslem Brotherhood, in the days when it was the crowning glory of the world.

Petrarch, by the way, declared that these same Church fathers were the greatest liars on earth.[359] But Petrarch might have been unduly prejudiced, for it is recorded that his sister was seduced by the holy (?) Pope John XXII., and he had witnessed some of the vile orgies of the Papal court at Avignon.[360] But anyone who reads Professor Draper's "Intellectual Development of Europe" will understand how much to believe of the declarations of the early professional Churchmen.

When Moslem Spain was the centre of an exalted civilization, the home of wealth, education and refinement, and the fountain-head of the material arts and sciences, Christian Europe was wallowing in dirt, degradation and barbarism, and inventing all sorts of falsehoods concerning Islam, prompted by their hatred and jealousy of a people who were immeasurably their superiors in every respect. This

---

[359] Francesco Petrarch, 1304-1374, Italian poet and humanist.

[360] Pope John XXII (Jacques Duèse), 1249-1334. Pontificate, 1316-1334. *New Catholic Encyclopedia*, 2nd ed., s.v. "John XXII, Pope."

spirit of animosity has descended from generation to generation, and its influence is plainly to be seen in nearly all the literature extant concerning Mohammed and Mohammedanism written by Christian authors. Every Christian author who has given the world a work depreciating the character of our Prophet and his teachings, has drawn his inspiration from the old and biased Christian sources, and has aided in perpetuating the falsehoods and false opinions of centuries ago. It is only when men like John Davenport and Godfrey Higgins dig under the rubbish of false history and bring to light some portions of the truth, that the English-speaking world gets glimpses of that eternal verity which rules the lives of so many millions of the human race.[361]

The Western mind is also prejudiced against Mohammedans and their religion by the newspaper and magazine contributions of the Christian missionaries who come to the East to by to convert the Asiatics. True, they meet with little or no success, but that is not due to a lack of zeal or earnestness. And yet you cannot justly blame these missionaries for misrepresenting you and your religion, for, in the majority of, cases, they, really believe they are telling the truth. They don't know any better, and their ignorance should be considered as mitigating their faults. As a rule, they are people of limited education and experience, and strong prejudices. They start with the firm conviction that they are right in their religious belief and that everybody else is wrong—hence their desire is to show the followers of other systems how far wrong they are. All they know of Mohammedanism they have acquired from the writings of ignorant Christians and, in their ignorance, they are disposed to magnify, distort and exaggerate the acts and expressions of Mohammedanism. But because they do so it is not a valid reason why you should despise and hate them—that would be diametrically opposed to the spirit of our Prophet's teachings—your duty is to be patient with them, to try to teach them the right way—to lead them out of the darkness of ignorance into the light of truth. The Holy Koran teaches us

---

[361] John Davenport, 1789-1877, author of *An Apology for Mohammed and the Koran* (1869).

that if we do this we will receive a great reward. Show them the true path, patiently and earnestly, and if they stubbornly refuse to walk in it, they will suffer and you will have the happy consciousness of having done your duty to God and your fellow-man.

And why is Islam the better way—the true and only way to salvation? Because it is founded upon that eternal truth which has been handed down to man from age to age, by the chosen Prophets of God from Moses to Mohammed, *Hazrat Mohammed Sallalaho Alehe Wassulim*. Because it is that eternal truth. Because it is the only system that will satisfy the longings of the soul for a higher existence. Because it is the only system known to man which is strictly in harmony with reason and science. Because it is free from degrading superstitions, and appeals directly to human rationality and intelligence. Because it makes every man individually responsible for every act he commits and every thought he thinks, and does not encourage him to sin by teaching him a vicarious atonement. Because it is elevating and refining in its tendencies, and develops the higher, nobler elements of humanity when it is faithfully, wisely and intelligently followed.

Now I am aware that this declaration will cause those Christians who hear it, to smile and ask me to go among the lower classes of Mohammedans and see if I can find any of the effects of the exalting, ennobling influences of Islam. I have been asked this question since I have been in India and I want to meet it right here.

If we are to judge a religious system by the moral and social character of the masses of its professed followers, Christianity will be so thoroughly condemned that it will speedily pass out of existence. Take a professed Mohammedan and compare him with a professed Christian of the same intellectual calibre, the same education and the same opportunities for obtaining secular knowledge and I am confident that the Mohammedan will show a cleaner moral record and higher spiritual perceptions than the Christian. Some of the most wretchedly degraded and fanatically superstitious people I have ever met in my life called themselves Christians. Of course they were not Christians, nor did they have even the remotest conception of the true teachings of Jesus of Nazareth, but they believed, or claimed to believe, in the Christian dogmas. Now every Mussulman knows that it requires something more to make a true Mohammedan than the parrot-like repetition of certain words or sentences. Because a man says he is a Mohammedan it does not follow that he comprehends or lives up to the spirit of the teachings of our Prophet; if he does not do so be cannot properly be taken as an example of the effects or tendencies of Islam. No religious system can fairly be judged by the acts and expressions of its professed followers; its fully established fundamental teachings and tenets only should guide us in forming an opinion of it.

And here let me assure you that there is nothing in the Islamic system that tends to immorality, impurity, social degradation, superstition nor fanaticism; on the contrary, it leads on to all that is purest and noblest in the human character, and when you see a professed Mohammedan, who is unclean in his person and habits, who is untruthful, cruel, intolerant, irreverent or fanatical, you may at once conclude that he is not a true follower of Islam and that he fails utterly to grasp the spirit of the religion he professes.

And now let us consider briefly some of the salient principles—the most prominent features of our religion, *viz.*; the Unity of God, the inspiration of our Prophet, cleanliness, prayer, fasting, fraternity, alms-giving, and pilgrimage. These may be said to form the foundation of the system; and I believe that an understanding of them is all that is necessary to commend the system to any intelligent person.

The truth of the Unity of God is apparent in all the manifestations of nature and has been clearly and unmistakably taught by the founder of every religion known to man. It is a historical fact that the Christian dogma of the Trinity was invented by the Bishop of Antioch 300 years after the death of Jesus, and that the latter never taught it and never heard of it.[362] Idolatry and a multiplicity of gods have

---

[362] Reference to Theophilus of Antioch, fl. 180, who was the first to use the word "trinity" in writing. Webb seems to have fused Theophilus with the establishment of the Nicene Creed a century later, which

invariably been the inventions of misguided men and such doctrines were never taught by truly inspired religions teachers.

Watch the course of nature if you are disposed to believe that it is the work of several gods. We see everything in the Universe moving regularly and systematically under the guidance of an unseen power. The planets revolve steadily and unswervingly in their orbits; the seasons come and go, the winds rise out of the unfathomed depths and go howling like demons through space or fan our heated brows as gently as a mother's kiss; the trees slowly rise out of the earth, spread their branches in cooling shade, or burst forth in a glorious bloom; the tiny plant expands until it can pour forth its tribute to the glory of God in fragrant blossoms; under fixed laws the fruits and grain appear as food for man and the grass and herbage for the animals; water falls from the clouds to revive and nourish the parched and thirsty earth; all the manifestations of terrestrial and celestial life and force are governed by a master hand that permits no change or variation in their general course.

How grandly and simply the Holy Koran expresses some of the signs of God!

"Verily God causeth the grain and the date-stone to break forth; He bringeth forth the living from the dead and the dead from the living. This is God!"

"He causeth the dawn to break, and hath ordained the night for rest, and the sun and the moon for computing time." "And it is He who hath ordained the stars for you, that ye may be guided thereby in the darknesses of the land and of the sea!"

"And it is He who sendeth down rain from Heaven; and We bring forth by it the germs of all the plants, and from them bring

---

formalized the trinity as a profession of Christian faith in 325. *New Catholic Encyclopedia*, 2nd ed., s.v. "Theophilus of Antioch" and "Nicene Creed."

We forth the green foliage and the close-growing grain; and from the palm-trees the low-hanging date-clusters out of their sheaths; and gardens of grapes, and the olive, and the pomegranite, like and unlike. Look, ye, on their fruits when they fruit and ripen. Truly, herein are signs unto people who believe."[363]

From the standpoint of human reason, which is the most rational conclusion—that there is one God or three or more whose power controls the Universe? Spiritual science demonstrates, beyond doubt, the Unity of God, while evidences of the truth of this doctrine greet us at every turn in our daily lives. These evidences will be convincing to any intelligent man who will throw aside his prejudices and consider them carefully.

What proof have we of the inspiration of our Prophet? The proofs are ample and abundant, but in order to present them fully and concisely I would need more time than is at my disposal now, and would be obliged to enter upon the discussion of the philosophic side of Islam, which is much broader and deeper than the average Mussulman imagines. But we really need no stronger proofs than the Holy Koran, the character of its and the Prophet's teachings, and the well-authenticated traditions and records of his life. No man could have lived as he did, taught as he did and accomplished what he did with no power animating and guiding him but the impulses of fallible human reason. One of the first things to impress me, after l began the study of Islam, was the persistent, steadfast earnestness of our Prophet in the face of obstacles and discouragements that would have utterly defeated an ordinary man. He evidently believed fully and perfectly what he taught. Had he been a man of ordinary mentality, vacillating and uncertain in his ways, and given to following visionary projects, his sincerity might have been less valuable as evidence; but we find that from boyhood to

---

[363] Qur'an, 6:95-99.

youth and manhood he was remarkably well-balanced mentally, and earnest, honest and rational in everything he under-
took. His moral character, too, was far above that of the average boy and youth of his time, and his steady, reliable
course of life won for him the love and esteem of all with
whom he came in contact. Wealth, power and an exalted social position were within his grasp, and he could have realized every human ambition and gratified every human desire
and impulse had he chosen to avail himself of the worldly
advantages offered to him. But he chose the better way; he
turned his back upon comfort and luxury, sacrificed all that
makes life dear to the average man, and suffered insult, persecution and extreme privation in order to teach the truth to
his fellow-men. These facts taken in connection with the
character of his teachings should be sufficient to convince
any man that be was truly inspired.

If you will carefully analyze those teachings you will
find that, in their ethical aspect, they are exactly in harmony
with the ethical teachings of Moses, Abraham, Jesus, and
every other truly inspired prophet known to history. The system he promulgated differs from that previously given to
man, because our Prophet's mission was to present a thor-
ough and complete code, the general purpose of which was
to correct the abuses and destroy the errors that had grown
about the doctrines taught by his prophetic predecessors. His
manifest purpose was to win mankind from idolatry, and to
present a series of rules or laws which, if followed faithfully
and intelligently, would draw men closer to God and make
them purer and cleaner, mentally as well as physically, and
better in every respect. He accomplished this mission fully,
and was not called hence until he had seen the Islamic system firmly established in the hearts and minds of his followers.

Of course, at this time, we can consider the Islamic
system only in its popular or esoteric aspect. As before
stated, it has a deeper, more philosophic aspect than is apparent at a first glance. But its chief beauty, viewed superfi-
cially, is its perfect adaptability to the spiritual needs of all
classes of humanity, from the humblest laborer to the most
advanced thinker and man of letters. There is nothing in it

that does violence to reason or common-sense or that is in any degree contrary to the natural instincts of justice and mercy. It requires no belief in the supernatural, nor the adoption of any absurd superstitions or impossible theories. Purity of thought, word and deed, perfect mental and physical cleanliness, and steady, unwavering aspiration to God, coupled with pure, unselfish, fraternal love, are the principal ends sought, and the means are as perfect as it is possible for man to conceive of.

Our Prophet forcibly declared that prayer was the corner-stone of religion, and he laid greater stress upon this than upon any other feature of his system. In order to show the solemnity and importance of prayer more plainly, as well as to secure the carrying out of another principle, the "Woozoo," or ablution was ordered. It was his evident intention to impress upon his followers the idea of cleanliness in such ways as were the most effective and permanent, and in the rule of Ablution, as well as in other rules, we readily see that he understood and appreciated the force of habit. No Mussulman who prays the required number of times daily ever thinks of. praying without thinking also of his "Woozoo," and thus he is reminded five times a day, at least, that he should have clean hands, face and feet, and in responding to the calls of nature he is cleanly to the last degree—far more so than the average man of any other faith. Thus he acquires habits of personal cleanliness which he cannot break away from without breaking away from his religion. All the evidence at hand, bearing upon the subject, tends to show that our Prophet not only intended that the hands, feet, face and private parts should be kept perfectly clean, but that all parts of the body and the clothing should be clean when we turn our face toward the Kaaba and our hearts toward God. In my opinion the Mussulman who prays in soiled clothing violates the spirit of the law, no matter how thoroughly and carefully he has performed his "Woozoo." I believe, also, that the floors of mosques and other places of prayer should be as nearly perfectly clean as it is possible to make them.

"But," someone may ask, "all Mussulmans are not so distressingly and painfully clean, are they?"

I regret to say that I have met some who were not, but I think that was because they failed to grasp he meaning of our Prophet's teachings. His apparent purpose was to foster and encourage cleanliness by the force of habit. When a man once gets into the habit of doing anything at a certain time his mind naturally reverts to that thing when the stated time arrives. To establish a good habit we usually commence to follow it from a sense of duty, and if it always comes to us as a duty it is much more liable to cling to us than if it is a matter of inclination. This is one of the points of the superiority of Islam over the so-called Christian system. In the latter the follower is taught that it is very meritorious to be clean and to pray, and that he ought to observe these rules—that is, of course, when he feels like doing so. If left to his own inclinations, man is liable to do those duties which are the easiest and most comfortable for him, or to his material, earthly advantage.

The same ideas may be applied to prayer. Our Prophet taught that it was indispensable. Suppose that he had said:

"Prayer is a good thing; you ought to pray; I would advise you to pray five times a day—if you can make it convenient. Pray often—whenever you can."

How much praying do you suppose there would be done to-day among the Mussulmans of the world? Very little I can assure you—just about as much as there is done among the masses of Christians to-day. The Bible says: "Pray without ceasing"[364] and in the most forcible terms exhorts to prayer and fasting; but in my long and extensive experience among orthodox Christians I have found many who do not pray at all, some who pray once a day,—when they retire at night and some who pray twice-when they retire at night and when they get out of bed in the morning. I have never met one, who was in the habit of praying oftener than twice a day and none who ever made his ablutions regularly before praying.

A professed Christian once said to me: "Oh yes that's all very well, but when a Christian prays he means it—it

---

[364] Thessalonians 5:17 (KJV)

comes from his heart, but when a Mussulman prays he does so mechanically and there is no meaning in his words."

I do not believe that that is the rule. There are thousands of Christians who repeat the Lord's Prayer and "Now I lay me down to sleep," without a thought of what the words mean, and just as mechanically and parrot-like as any Mussulman ever did. But the Mussulman has an advantage even if he does repeat the words of his prayer mechanically; the prostrations and genuflections keep him constantly reminded of his duty and prevent his thoughts from wandering off elsewhere. If he prays five times a day he must, of necessity, get as much soul into these five prayers as the Christian does in one short one, and I am inclined to believe that he accomplishes considerably more.

But if the Mussulman prays listlessly and indifferently it is because he has failed to comprehend the Prophet's teachings, and not because the system is faulty. We are clearly taught that prayer, to be effective, must proceed from the heart, and one of the purposes of the ablution is to prepare the mind to throw off its worldly concerns and concentrate itself wholly upon the words spoken and the prostrations and genuflections performed. We are to follow the idea of presenting ourselves before God physically and mentally clean, and to address Him soulfully and reverently. Those who have read the words of Imam Al Ghazzali can discern in prayer a vast deal more than empty words and in the Islamic system a most effective means of leading people into the habit of prayer, even when they are not induced to pray by their own inclinations.[365]

A Mussulman told me once that he was rather inclined to believe, sometimes, that it was not necessary to follow strictly the methods of prayer followed in the days of our Prophet and his immediate successors; and when I asked him his reason for thinking so, he said it was because the Arabs were not naturally a praying people, and some hard and fast rule was necessary in order to get them into the habit of

---

[365] Abu Hamid Muhammad ibn Muhammad al-Tusi al-Ghazali, 1058-1111, noted Muslim theologian, jurist, mystic and reformer. *Encyclopaedia of Islam*, New ed., s.v. "Al-Ghazali."

praying. I called his attention to the conditions prevalent among those who have adopted European ideas and customs, and convinced him, I think, that there was just as much of a necessity for a hard and fast rule to-day as there was 1,200 years ago. If those ideas, habits and customs, which we group under the name of 19th century civilization, are calculated to produce and encourage a disposition to pray I have failed utterly to discover it, and I have lived within that civilization all my life.

Man is a creature of habit and, as a rule, when he once drops into a groove he rarely gets out of it without an unusual effort, unless he does so in order to follow something a little nearer to the earth. If he acquires the habit of praying five times a day it will cling to him until he dies, and his prayer will increase in earnestness and soulfulness as his knowledge of the fundamental principles of his religion increases.

One of the wise provisions of the Islamic system is the rule relative to congregational prayer. The Mussulman is taught that he should always pray in company with others when it is possible for him to do so. Now there are several very good and sufficient reasons for this rule which can only be discussed under the head of Islamic Philosophy; we can only look, now, at its outward aspect. In the first place it is calculated to break down caste distinctions and place the servant and his master upon a common level before God, in whose presence all men are equal. The beggar, the merchant, the shopkeeper, mechanic, millionaire, the professional man all stand elbow to elbow as brothers in the mosque at the time of prayer, and when a number of Mussulmans come together anywhere at the stated hours they are, in duty, bound to cast aside social distinctions and pray together. It is also the duty of the Mussulman to pray wherever he may happen to be when the hour of prayer arrives; or, if the place is unsuitable, to seek a better one. It is a severe blow to the pride of a young man when he must kneel in the presence of his non-praying companions or retire from their company for the avowed purpose of praying.

In fact, the tendency of all the rules relating to prayer is to secure honesty and frankness of purpose, complete devotion to religion, and soulful, hearty worship of the one true God. The word Islam means, literally, resignation to the will of God, and hence prayer, in its purest, most perfect aspect, must be the very essence of the system.

It is generally admitted that a man may call himself a Mussulman if he simply declares his belief in the Unity of God and the inspiration of the Prophet, but he certainly cannot be called a true follower of Islam unless he prays from the very depths of his heart and makes the purpose of his prayer to bring his soul nearer to God.

Probably the most curious and absurd objection I have ever heard against Islam was in connection with the subject of prayer. I was talking with an Englishman upon the general subject of Mohammedanism and after I had replied to a number of objections he said:

"Well, there is one thing about Mohammedans that would prevent my becoming one; and that is that every one of them is duck-legged and has knees in his trousers from sitting on his legs and feet so much in prayer."

After I had, to some extent, recovered from my surprise I replied: "In my opinion, it is very much more to one's advantage to go into Paradise duck-legged and with knees in his trousers than not to get in at all."

And since that time whenever I see a Mussulman who is duck-legged and has knees in his trousers I, at once, have a feeling of respect for him for I know that he is attending to his religious duties.

The subject of prayer leads us, quite naturally, to that of fasting, which is one of the most important features of the Islamic system. In the doctrines of every religion, fasting appears as one of the prominent articles of discipline. Jesus taught it repeatedly and emphatically and insisted upon it as one of the first necessities of spiritual development, but the designers and architects of the so-called Christian system failed to present it as a fixed and inviolable law and hence it has fallen into desuetude, particularly among the Protestants. Some Roman Catholics still make a pretence of fasting but I never heard of a Protestant who fasted.

The Holy Koran says:

"O Believers! a fast is prescribed to you, as it was prescribed to those before you, that ye may fear God."[366]

This is a fixed law and it is obeyed, with more or less sincerity, in every Mohammedan community on earth, during the month of Ramzan in every year. Whether, in these degenerate days, the results are what the fast is intended to produce may be questioned in many cases, but there are, undoubtedly, many other cases in which a vast amount of spiritual good is attained.

From the fact that every inspired religious teacher, from the beginning of history to the time of our Prophet, taught his followers to fast, and that all those who have attained to high spiritual conditions in any part of the world, have urged the same practice, it is reasonable to infer that there is some valid reason for it. It is a well-established fact that the lower bodily health and strength become, the more vivid and intense become the psychic powers. Anyone who will try the experiment will find that a fast of twenty-four hours will increase his sensitiveness very perceptibly and also make his mental powers brighter and more active. We can think with greater facility and control our thoughts more easily when our stomachs are empty than when we have eaten a hearty meal.

In view of all the discoveries of materialistic science it is quite plain that there is something more in fasting than a meaningless form, a foolish superstition, or even a method of discipline. Its true purpose is to relieve the soul of the burden of earthly desires and thoughts and prepare it for the reception of spiritual truth. Or we may say that it is the cleansing and purifying of the soul so that it may be in a fit condition to appear before that higher spirit which dwells within every man.

But at this time we have only to do with the disciplinary results of the fast, which are the most beneficial to those who perform this duty solemnly and sincerely. Even if it is performed mechanically, and without any idea of its higher purposes, it takes the Mussulman a little further along the road toward spiritual perfection than he would go if he made

---

[366] Qur'an, 2:183.

258

no attempt to fast. Of course, the more earnest and sincere the faster is, the higher will be the results obtained and the greater will be the reward hereafter; the man is left to choose for himself the course he will pursue knowing, at the same time, that the law requires him to fast. Obedience to God is meritorious, but earnest, cheerful obedience is far better than reluctant and indifferent compliance.

Generally speaking the fast gives one month in the year to God, as one day in the week and five times in the day are given to Him. During this month the Mussulman is not only required to refrain from food from dawn to sunset of each day, but the fast is considered broken and ineffectual if at any time during that period he gives way to anger or any passion or indulges in slander or falsehood. This curbing of the propensities of the lower nature is just as much a part of the fast as refraining from taking food. Now is it possible for a man to be virtuous for one month in each year without having his general character benefited to some extent during the other eleven months?

But the real purpose of the fast is higher than the mere temporary discipline of the lower or animal nature; it is a wise provision which, if followed intelligently, sincerely and prayerfully, will inevitably bring about a closer communion between the soul and that spirit of God which the Holy Koran tells us is nearer to man than his jugular vein.

The Holy Koran says:

"Verily there is pardon and a great reward for those who fear their Lord in secret. And be your converse hidden or open, He verily knoweth the inmost recess of your breasts."[367]

Now if this is true the purposes and efficacy of fasting and prayer or meditation are at once apparent. The abstaining from food brings the soul forces into ascendency over the animal inclinations and the soul is voluntarily presented and laid open to God. Can anyone doubt that great good must come from such an exercise?

Our esteemed brother Hassan Ali, the earnest and faithful Mohammedan missionary, has written an essay on

---

[367] Qur'an, 67:12-13.

259

Fasting which contains a number of valuable ideas and suggestions. It is printed in Urdu and should be in the hands of every thoughtful Mussulman in India.

It was apparently our Prophet's purpose to induce his followers to acquire the habit of observing the fast as well as the habits of prayer and cleanliness, realizing that these habits would descend to their children and their children's children from generation to generation, thus keeping mankind in the true path even if all did not walk closely in it. Twelve hundred years have passed and the month of Ramzan is looked forward to and observed by the masses of Mohammedans throughout the world. They have acquired the habit and it clings to them, doing more good than the masses have any idea of.

No one who has given the Islamic system careful and unprejudiced study can fail to see how wisely and judiciously it was constructed and how broad and effective its results must be if it is followed intelligently and faithfully. We all know the force of habit, good or bad, and how easy it is to drift into vice and sin. It seems very much easier and more comfortable to be wicked than to be good in this generation.

The poet has truly said:

"Vice is a monster of such frightful mien,
That to be hated needs but to be seen,
But seen too oft, familiar with its face,
We first endure, then pity, then embrace."

If we once begin to stray away from the path of religious habit and duty it is very easy for us to drift with the tide of worldliness out into the broad and tempestuous ocean of sin and despair. Hence the wisest course for us to pursue is to study our religion until we arrive at an intelligent comprehension of its principles and purposes, and then to follow its forms and duties earnestly, honestly and sincerely.

And now let us glance at another of the most important features of the Islamic system, viz., fraternity. One of the first acts of our Prophet after he had taken up his abode in Medina was to commence the organization of the Moslem

Brotherhood; that fraternity of earnest, loyal souls which was destined to create a grand and glorious revolution throughout the entire East. Through all his life and teachings we find a strong sentiment of brotherly love and devotion running like a thread of silver through a field of gold; it permeated all the thoughts and actions of the early Moslems and was a potent factor in bringing about their grand and glorious achievements at arms.

A story is told of an old man who, on his death-bed called his sons to him to receive his last blessing. When they had assembled and stood in tearful attention waiting to hear his words, he told one of them to bring a number of small sticks. The son obeyed and in obedience to his father's command broke one of the sticks with his fingers. His father then told him to put all the sticks together and try to break them, but he found that he could not do so although he brought all his strength to bear upon them.

"You see," said the old man feebly, "how easily you broke the stick when it was away from the others and how difficult it is to break them when they are all together. Herein is a lesson which I want you to keep in mind after I am gone. In union there is strength."

This was apparently the lesson which our Prophet taught his followers, knowing that while they were united in a bond of true fraternity and inspired by sincere devotion to God, they would be invincible. And so long as that spirit of fraternity existed in the Moslem ranks, so long as the fire of brotherly love and devotion burned in the hearts of the defenders of the true faith, so long did Islam continue on its irresistible march toward the zenith of its glory and power. But just as soon as dissension and disunion made their appearance the power of the Moslem arms began to weaken, and the moving column found itself utterly unable to overcome the obstructions that lay in its path. Compare the condition of the Moslem world in the eighth century with what it is to-day and see if you cannot find therein a lesson of great value.

The plain duty of every Mussulman who loves his religion and reveres the memory of our Holy Prophet is to endeavor earnestly and persistently to revive that sentiment of fraternal love which was so prominent in the character of our brethren who lived and loved and worshipped the one true God 1,200 years ago. Those who follow Islam should be brothers before God, and should cultivate within their hearts an unselfish, fraternal love for each other which should influence their contact and communication with each other throughout their lives. To do this mutual forgiveness and mutual concessions are necessary—let us bear with each other's faults and weaknesses, praying to God for patience and strength and clasping hands honestly and fervently as true brothers should.

There is, there can be, but one true religion although all the numerous and varied systems known to man may have more or less of the one truth underlying them. By careful and unprejudiced study we can readily trace this truth through all the teachings of the prophets. By weeding out the false translations, and interpolations in the New Testament of the Christians we find it plainly taught by Jesus of Nazareth, but he died before his mission was accomplished and failed to formulate a system of practice for the purpose of firmly establishing the truth in the hearts of his followers. It appears quite plain that the majority of his twelve disciples did not understand him and failed utterly to grasp the spirit of his teachings. The present system called Christianity was really built upon the teachings of Paul, about three centuries after the death of Jesus; and Paul not only never saw the Prophet of Nazareth but seems to have had only a very vague idea of what the latter taught.

Our Prophet, the last and greatest of all the prophets, not only taught this great truth to his immediate followers, so that they evidently understood it, but he formulated and firmly established a complete and perfect system of practice calculated to impress it deeply on the hearts of the Arabs and to carry it from generation to generation to all nations in all ages of the world. His mission was fully accomplished before God called him to his reward, and the legacy of truth he left to sin-stricken humanity is the grandest monument that can be conceived in commemoration of his lofty, stainless character and his noble self-sacrifice. It is the plain and unmistakable duty of every Mussulman who loves and reveres the memory of our grand Prophet to use his earnest efforts to assist in planting that monument—Islam—upon the soil of every nation.

It is an old and true saying that, "Truth crushed to earth will rise again." Islam has passed through various stages during the past twelve centuries and, until within a comparatively few years ago, it seemed to be steadily declining under the crushing weight of a grossly materialistic civilization. But there are signs of a great change, of a revival of that old vigor which made it the glory of the world. Its progress in Africa during the past decade has been phenomenal; it has found a firm foothold in Europe, and with God's help we propose to establish it in liberal progressive America where, I feel confident, the masses of people are waiting to receive it. With all these evidences before you, can you doubt the final result? "*La illaha illala. Mohammed rasoul Allah.*"[368]

---

[368] Translates from Arabic as "There is no God but Allah and Mohammed is his messenger."

YANKEE MUSLIM, ED. BY BRENT D. SINGLETON

APPENDIX A.

WEBB INTERVIEW IN THE *MOHAMMADAN OBSERVER*

When I went to see Mr. Webb about the reasons which led him to embrace the Mohammadan religion. I found him seated in a chair and dressed in the Mohammadan costume. Mr. Webb, who is very polite, good-humoured, and well-educated, received me cordially, and when asked if he had any objection to being "interviewed" and having his views published, replied that he would be very pleased to afford any information to the public and to answer any question which any one cared to ask him. He spoke of the Christian religion in general in a most harsh way, as he thought it was incomparable with the dogmas of the Mohammadan faith. His opinion of the Bible was very poor, and when questioned as to whether there were not likely to be as many defects in the Koran as in the Bible, replied that the Koran was the result of a combination of facts actually recorded at the time; while the Bible record was one produced by the gathering together of a number of documents and praying for the right one to leap up. Speaking of the Salvation Army, Mr. Webb thought nothing of them at all. These are all Mr. Webb's opinions, and I give below exactly what he has to say on the subject—

Q.—Mr. Webb, what Church did you belong to before you embraced the Mohammadan faith?

A.—The Presbyterian Church. I was born at Hudson, New York. When I was young I went to Sunday school. In those days I was a pretty wild kind of a boy, and I used to go to church simply for the sake of seeing nice-looking girls and escorting them home. It was a most delightful task. When I was a little older, I changed from the Presbyterian church, to the Episcopal one, as there were equally nice young ladies to look at in the latter church.

Q.—Then you only went to church to look at young ladies?

A.—Yes, I gave religion no thought. About the age of 20 I met a family of Universalists, and they began to preach religion to me, and I began to think of religion. The more I thought of it, the more absurd it seemed to me. Now take the mystery of the Trinity.

Q.—Were you not taught the mystery of the Trinity?

A.—Yes, about 20 years ago, but I was unable to find anybody who could enlighten me.

Q.—Did you interview the clergy about it?

A.—Yes. They told me that I must not go into these things, as it was the mystery of God; and I told them that if the Christian religion was necessary for my well-being I ought to know something about it.

Q.—Have you read through the Bible?

A.—I have studied the Bible. I do not believe in the Immaculate Conception and the Vicarious Atonement.

Q.—Why don't you believe in the Atonement?

A.—Because I could lead the most wild life, and with 15 minutes' atonement expect to go the heaven.

Q.—What do you mean?

A.—By simply declaring my belief in Jesus Christ.

Q.—What do you think of the High Church?

A.—There is no Christianity in the High Church.

Q.—What do you again mean?

A.—Because it does not teach religion. The whole of the Catholic Church systems and forms of ceremonies are based upon the old Pagan system of ceremonial magic.

Q.—Are you speaking from what you have read, or from actual observation?

A.—I have been deeply interested in these studies for years. I have never lost an opportunity of interviewing anybody at all, be the person never so high. I have studied books, and have interviewed the clergies.

Q.—How long have you embraced the Islam faith?

A.—For the last 5 years.

Q.—Are you married?

A.—Yes, and my wife has also embraced the Islam faith.

Q.—What do you think of the Roman Catholic Church, as compared with the Protestant Church?

A.—Well, the Protestant Church allows more liberty, where as the other Church does not.

Q.—But is it nice to allow the ignorant to understand more than they need for purposes of salvation?

A.—They should be allowed freedom. The Roman Catholics are never allowed to read the Bible.

A.—But I have seen a Bible in a Roman Catholic house.

A.—You may have; but it is against the rules.

Q.—But you believed in the Christian religion once?

A.—No, never in my life.

Q.—Don't you think the Mohammadan who has embraced the Christian religion is likely to have the same opinion of the Islam faith that you have of the Christian faith?

A.—Yes, there are a great many who do say the same thing against the faith they have ignored. I remember an article written in reference to Mohammadan marriage; but it was all false.

Q.—Do you believe in polygamy?

A.—Yes, I was speaking to the American Consul General in Calcutta, and he was horrified that I had turned Mohammadan. He asked me whether I believed in Mohammad as well as I believed in Jesus Christ; and I said that I most positively did so. Well, he then said that Jesus had not half a dozen or more wives; and I said that Mohammad had not either. People have to be educated to polygamy. It is not obligatory at all. It is from childhood that one must be taught the advantages or otherwise of polygamy. In America there are 90 per cent polygamists.

Q.—Now what really induced you to turn Mohammadan?

A.—I began to be interested in it about 20 years ago. I am near 45 years old. For nine years I had no religion. I followed the old materialistic idea. By reading Bulwer Lytton's "Zanoni," my curiosity was excited. My association with Christian clergymen and others had convinced me that the masses of Christians neither knew nor cared to know what their system was. The priests were as much in the dark; they shrank from investigating anything which promised to deprive them of their valuable hood.

Q.—What do you think of the Greek Church?

A.—I can't say.

Q.—Don't you think it is the primitive Church?

A.—Yes, it may have been the primitive Christian Church.

Q.—Do you believe in the Deluge?

A.—No, not as an actual fact.

Q.—What do you intend doing in America?

A.—We intend to establish a high class weekly newspaper, to be devoted to the real doctrines of Islam. We propose to establish a place for the issue of pamphlets and books, to establish a free library and reading room for the masses, and a free lecture-room where lectures will be delivered once or twice a week. We propose to educate the English speaking people, and to overturn the false impressions that have been made by many writers. Also to establish in the various cities branch societies to propagate the Islam Faith.

Q.—Do you think there is a possibility of converting Americans?

A.—Yes, most certainly.

Q.—What makes you think so?

A.—Because during the last 10 or 15 years there has been a great falling off in confidence in the Christian faith. People have been drifting away from it, and hence there are so many societies formed, such as the Sectarian Society, the Theosophical Society, and various others; and people are anxious to investigate, in order to get to the bottom of the system of religion. Now in St. Louis, where there is a population of half a million inhabitants, where a special staff of reporters were sent to take the census of the people who attended church, it was reported that only 7,000 people did so out of the half million; and rest attended the various places of amusement.

Q.—But that is no criterion. A man's own house may be as good as his church. He may say his prayers at home, and should not be classed as one who preferred the saloon to the church.

A.—Yes, but in America things are different. There is no such thing as saying prayers in one's own house.

Q.—Don't you think there are as many defects in the Koran as there are alleged to be in the Bible.

A.—No.

Q.—Why?

A.—The Koran is a thing which I should not like to discuss in a newspaper article. But Mohammad's words were actually recorded, while Jesus's words were not recorded. The present Bible was complied 300 years after Christ died. The Bible was compiled from a collection of mystical books, by a set of people who prayed that the true Scriptures would leap from the pile.

Q.—That is only a matter of opinion. Where did you find that?

A.—In books.

Q.—Do you believe in Sunday?

A.—Yes, only as a day of rest, when needed.

Q.—But the Mohammadans have a day also?

A.—Yes, but the Sunday of the Christians is really the Saturday of the Jews.

Q.—What do you propose doing here in Bombay?

A.—Nothing. I am neither a missionary nor a theological Juggler. I have only come down here to make the acquaintance of my Mohammadan friends, and not to convert people.

Mr. Webb, at the end of the interview, expressed himself very pleased at the opportunity he had had of speaking on the subject; and stated that he would be very pleased indeed to discuss the question of the Christian religion, as compared with the Islam faith, and to hear any arguments that people cared to advance on either side of the question.

APPENDIX B.

ALEX R. WEBB STATEMENT OF AGE, EDUCATION, EMPLOYMENTS, ETC.

Mr. Alexander R. Webb was born on the 20th day of November 1846 at Hudson, Columbia Co., New York, his father being editor and proprietor of the Hudson Daily Star which paper he conducted for about twenty-five years. The first fourteen years of Mr. Webb's life were spent at Hudson the foundation of his education being laid at two private schools. He was then sent to Warner's Home School at Glendale Mass. where he remained about three years after which he entered Claverack College. At the conclusion of his course here he returned to his home in Hudson and at the suggestion of friends decided to learn a trade. An arrangement was made with a friend of his father Chas. E. Butler a jeweler at Hudson, and he spent about two years and a half without remuneration learning the Watchmaker's Trade and acting as salesman. Desiring to seek his fortune in the West he adopted the suggestion of a friend of the family and went to Chicago where he was subsequently engaged as salesman in the jewelry house of Giles Bro. and Co. In 1869 he married a daughter of L.W. Conger of Chicago and formed a partnership with him in the jewelry business at the corner of Clark and Madison streets. They were burned out in the great fire of 1870 and Mr. Webb returned to New York where his father had secured a position for him in the jewelry house of Tiffany and Co. on Union Square. He remained here until 1873 when owing to his wife's ill-health and in response to the solicitation of her parents he returned to Chicago taking with him a strong letter of commendation from Tiffany and Co. After a short engagement with the jewelry house of Wendell and Hyman in Chicago his father-in-law who had opened a bank at Unionville Putnam Co. Mo. purchased a half-interest in the Unionville *Republican* and offered it to Mr. Webb if he would come and take it. Having always manifested a taste for literary pursuits he yielded to his desire to become a journalist and accepted the offer. He had never taken more than a superficial interest in politics and soon found that his ideas were too strongly Democratic to admit of his running a Republican paper to the satisfaction of the local leaders of that party and as he had acquired entire control of the paper he followed the People's movement in 1874. In 1876 he sold out, gave up all active participation in politics and became City Editor of the St. Joseph *Gazette*. He remained on that paper but three months when he accepted an offer to go to St. Louis and take a position on the Morning *Journal*. The proprietor of that paper, however, failing to pay salaries promptly he took a position as salesman with the jewelry firm of Eugene Jaccard and Co. Subsequently the proprietors of the *Journal* purchased the Evening *Dispatch* and a satisfactory adjustment of the salary question having been arrived at he became City Editor of both papers. The *Dispatch* franchise was afterward purchased by Joseph Pulitzer and the *Journal* was consolidated with the *Times* Mr. Webb taking a position as special on the local staff of the *Times-Journal*. He subsequently accepted a position on the *Globe-Democrat* and afterward returned to the *Times*. He was with the latter paper when it was absorbed by the *Missouri Republican* and with all its employees was compelled to find other employment. He procured a position as salesman with Mermod, Jaccard and Co. jewelers. On the 1st of November 1883 he returned to journalism once again taking a position on the editorial staff of the Missouri *Republican* where he is engaged at present.

In January 1884 he became interested in the Oriental religions to the study of which with kindred philosophical and metaphysical subjects, he has devoted nearly all his leisure time since. This led him into other branches of Asiatic research and he became a member of a fraternal organization

273

which now has branches in nearly every city of importance in India. He opened a correspondence with natives of the higher castes and has formed connections of a character which convince him that he would be most cordially received in any thickly populated port of Asia under English rule. This connection he is also satisfied would aid him in the transaction of Consular business as he would be in a position to readily obtain information which it would be exceedingly difficult, if not impossible for an American or Englishman to procure by ordinary methods. His knowledge of Oriental peoples and customs is extensive and those who are intimately acquainted with him are confident that he would not only transact the business of the Consulate at Singapore faithfully and efficiently but that he would be a most creditable representative of this country.

He uses neither wines, beer, nor tobacco nor stimulants nor narcotics of any sort and his physical health is as perfect as anyone could desire. He comes from a long-lived family and his regular habits, systematic course of diet, composed mainly of vegetables and fruits, coupled with a naturally vigorous constitution render him capable of enduring extremely severe physical strains. He has not been ill enough to go to bed nor to require the services of a physician for more than twenty-five years.

APPENDIX C.

LETTERS BETWEEN WEBB AND MIRZA GHULAM AHMAD

Webb to Ahmad, 1886
(translated from the Urdu original)

Gentleman,

I have recently read your letter in a newspaper [name of newspaper unclear, however it mentions a Mr. Scott] which refers to your invitation to the Truth. I found myself interested in this movement. I have studied quite a bit about Buddhism, Brahmanism and also about Confucianism and Zoroastrianism, but not as much about Prophet Muhammad.

As for the straight path, I have been and am still with a Christian group and I lead a Church. I am not capable of teaching much more than basic morality.

I am basically in search of the Truth and feel sincere toward you.

Your Servant

Alexander Webb
3021 Easton Ave.
St. Louis. Missouri

Webb to Ahmad
February 24th, 1887
(Ahmad's reply to Webb's initial inquiry is unavailable)

Esteemed Sir,

I cannot adequately express to you my gratitude for the letter received from you under date of December 17. I had almost given up all hope of receiving a reply but the contents of the letter and circulars fully repaid me for the delay. I hardly know what to say in reply except that I am still very anxious to gain more of the truth than I have thus far found. After reading your circulars an idea occurred to me which I will present to you for your consideration knowing or rather feeling confident that you who are so much more spiritual than I, so much nearer to God, will answer me in a way that will be for the best. Were it possible for me to visit India I would do so only too gladly. But I am so situated that it seems almost an impossibility, I am married and have three children. For nearly two years I have been living a life of celibacy and shall continue to do so as long as I live. My income is not sufficient to justify me in giving up my business as it requires all that I can make to support my family; therefore, even if I had sufficient means to enable me to make the journey to India I would not be able to furnish support for my family during my absence. Therefore a visit to India being out of the question it occurred to me that I might through your aid assist in spreading the truth here. If, as you say the Muhammadan is the only true religion why could I not act as its Apostle or promulgator in America. My opportunities for doing so seem to me very good if I had some one to lead me aright at first. I have been led to believe that not only Muhammad but also Jesus, Gautama Buddha, Zoraster and many others taught the truth, that we should, however, worship God and not men. If I could know what Muhammad really taught that was superior to the teachings of others, I could then be in a position to defend and promulgate the Muhammadan religion above all others. But the little I do know of his teachings is not sufficient for me to do effective work with. The attention of the American people is being quite generally attracted to the oriental religions but Buddhism seems to be foremost in their investigations. The public mind, I think is now more than ever fitted to receive Muhammadanism as well as Buddhism and it may be that through you it is to be introduced in my country. I am convinced that you are very much in earnest, I have no reason to doubt that you are inspired by God to spread the light of truth; therefore I would be happy to know more of your teachings and to hear further from you. God, who can read all hearts, knows that I am seeking for the truth that I am ready and eager to embrace it wherever I can find it. If you can lead me into its blessed light you will find me not only a willing pupil but an anxious one. I have been seeking now for three years and have found a great deal. God has blessed me abundantly and I want to do His work earnestly and faithfully. How to do it is what has moved me—how to do it so that the most good may be accomplished. I pray to Him that the way may be pointed out clearly to me so that I may not go astray. If you can help me I hope that you will do so. I shall keep your letter and prize it highly. I will get the circulars printed in one of the leading American newspapers so that they will have a widespread circulation and I will send you a copy of the paper. They may reach the eyes of many who will become interested. I shall be happy to receive from you at any time matter which you may have for general circulation and if you should see fit to use my services to further the aims of truth in the country they will be freely at you disposal provided, of course, that I am capable of receiving your ideas and that they convince me of their truth. I am already well satisfied that Muhammad taught the truth that he pointed out the way to salvation and that those who follow His teachings will attain to a condition of eternal bliss. But did not Jesus Christ also teach the way? Now suppose I should follow the way pointed by Jesus, would not my salvation be as perfectly assured as if I followed Islam? I ask with a desire to know that truth and not to dispute or argue. I am seeking the truth not to defend my theory, I think I understand you to be a follower of the esoteric teachings of Muhammad and not what is known to the masses of the people

as Muhamaddanism; that you recognize the truths that under-
lie all religions and not their esoteric features which have
been added by men. I too regret very much that I cannot understand your language nor you mine; for I feel quite sure
that you could tell many things which I much desire to know.
However I am impressed to believe that God will provide a
way if I try to deserve His love. Blessed be His holy name
and I hope that I may hear from you and that we again may
some day meet in spirit even if we cannot meet in the body.
May the peace of God be with you and with those who listen
to your words. I pray that all your hopes and plans may be
realized. With reverence and esteem.

I am,

Yours Respectfully,

Alex R. Webb

Ahmad to Webb
(undated reply to above letter)

Dear Sir,
        I received your letter, dated 24th of February 1887
which proved itself to be great delight to my heart and a satisfaction to my anxieties. The contents of the letter not only
increased my love towards you but led me to the hope of a
partial realization of the object which I have in view—for
which I have dedicated the whole of my life viz, not to con-
fine the spread of the light of truth to the oriental world but,
as far as it lies in my power to further it in Europe, America
& Co. where the attention of the people has not been sufficiently attracted towards a proper understanding of the teachings of Islam. Therefore I consider it an honour to comply
with your request; and have a strong confidence in the Almighty Creator, Who is with me, that He will assist me in
giving you a perfect and permanent satisfaction. I give you
word that in the course of about five months I will compile a
work containing a short sketch of the teachings of the Al-
Quran, have it translated into English and printed and then

278

send a copy of it to you. I strongly hope that it will bring full and final conviction to a justful, considerate and uncontaminated mind like yours, ennoble your soul, endow you with a firm belief in God and improve your knowledge of Him. But perhaps it may be, that the various demands on my time may not allow me to spare a sufficient time for sending the whole work at once, in such a case I will send it to you in two or three batches. I will not end the communication of instruction to you by this treatise but will continue satisfying your thirst after the investigation of truth for the rest of my life. Your friendly words permit me to entertain the happy idea that I will in a short time have the intelligence that the instinctive moral greatness has directed not only to you but to many other virtuous men of America to the right way of salvation pointed out by Islam. Here I end my letter of earnestness and sincerity. May you and I be kept secure from all earthly and heavenly misfortunes and have all our hopes and plans realized.

Yours sincerely,

Mirza Ghulam Ahmad,
Chief of Qadian,
Gurdaspur District, Punjab,
India.

APPENDIX D.

FUNDRAISING LETTER FROM WEBB

[undated 1894]*

I have given up all my worldly advantages and find myself without the means of support. God has not allowed me nor my family to suffer. This vast country, four times larger than all India, is buried in the darkness of materialism, and many of its people are nominally following that false system known as Christianity. Yet there are many thousands of honest, earnest souls who are sincerely seeking for the light of truth.

Send me all you can spare from your own money and then urge your brother Mussulmans to do the same.

If you can send 1 rupee or 1 piaster it will help us. I f you can send me 100, 500, or 1,000, so much the better for you and for Islam.

The Christians spend millions of dollars every year to spread their false religion in the Orient; why cannot the Mussulmans spend a few thousands to spread the true faith here?

There was a time when Islam was the glory of the world; it can be made the religion of the whole world if each Mussulman in the East will do his duty to God and himself.

Help me to plant the standard of Islam in the United States of America, and millions of Americans will clasp your hands in fraternal love.

---

* Originally published in: "Nefeesa Keep Breakfasts," *New York Times*, July 16, 1894, p. 1

APPENDIX E.

LETTER FROM WEBB (IN MANILA) TO BUDRUDIN ABDULLA KUR (BOMBAY)

July 4, 1892[*]

My dear Brother,

Your letter, dated 1st ultimo and enclosing one from our esteemed brother, Abdulla Arab, is at hand. It gives me great
pleasure to know that you are taking such an active interest
in the proposed American mission, which, I feel assured now
will be a pronounced success. I am fully satisfied that when
the Mohammadans of India thoroughly understand the nature
of the movement and the character of the field in which it is
to be established, together with its possible effects upon the
propaganda in Europe, and upon the Moslem organization in
the East, they will give it their most earnest and active support. To me, it is in no sense an experiment likely to result in
failure, for I know the general tendency of thought in my
country and the general characteristic of my countrymen too
well to entertain for a moment the idea that anything but success is possible. Besides I have faith in the power and wisdom of Almighty god (Praised be His name for ever), and as
Islam is the true religion I feel confident that He will guide,
direct and support a movement for its propagation which

---

[*] Reprinted from: Alexander Russell Webb, *Lectures on Islam: Delivered
at Different Places in India.* Lahore: Mohammadan Tract and Book
Depot, 1893, 16-18.

promised such great results as this. I have for several years been convinced that there were unseen influences at work bringing about a condition of things calculated to overthrow the current erroneous religious systems of the world and establish mankind in one true system. But which that system was to be was to me uncertain until I arrived at a comprehension of the character and doctrines of Islam. I have had some strange experiences of which I hope to have the privilege of talking with you soon, and which have seemed to me as evidence that God was guiding me for some great and wise purpose, the ultimate object of which was the spiritual benefit of mankind. I do not desire to give you the impression that I believe I have been inspired directly, but circumstances have so shaped themselves in my life that they have drawn me directly toward the movement in which I am about to engage with all the earnestness, vigour, and intellectual ability that God has given me. As I wrote to you more than a year ago it seemed that I was destined to work out quietly, and in my own way the bringing of my countrymen to a knowledge of Islam, and it hardly seemed probable then that any other way was open. But God in his wisdom has opened a broader, surer, and better way generally, and I can see clearly how I am to devote my efforts to bring about more speedily and thoroughly the desired results. I am impatient to meet you and talk freely with you concerning the matter, for I feel sure that you will agree with me not only that the object is of the grandest importance, but that there has been something more than ordinary human agencies at work in bringing the project to its present condition. God is great, and will surely guide his servants in the right way.

I have sent my resignation to my Government, and I am awaiting instructions as to the disposition of the office and effects. If I am obliged to remain here until my successor arrives, I will be unable to leave before about the 1st of October, but if I am authorized to turn the office over to some one here I can leave the latter part of August or early in September. I will then start for Bombay *via* Singapore, Rangoon, and Calcutta. My wife and family left for the United States on the 8th of last month, and will got to a fruit ranch about 40 miles from San Francisco, where they will remain until I am fairly settled in New York.

The books you sent me have been a source of the deepest interest to me, and I have learned a great deal that was new to me, concerning the diversity and general direction of Mussulman thought and opinion. I understand that you do not endorse all the views and the opinions expressed and judging from your marginal notes, I am inclined to the belief that you and I will not be very far apart in our views. I have, or think I have, a good conception of the spirit of Mohammad's teachings and the fundamental principles of Islam. Hence I am force to reject many of the opinions I have encountered. Hughes' *Dictionary of Islam* abounds with prejudice and error, and there are some suggestions made by Syed Ameer Ali that I can hardly endorse. The *Critical Exposition of the Jehad* is a most important and interesting work to me, but it ought to be rewritten and revised, so that its English will be smoother and more concise. But in his treatment of the subject the author has furnished me with matter which will be of inestimable value to me when I begin active work in America. But of these matters I hope, if it is God's pleasure, to talk with you when we meet. God grant that the hearts and minds of our brethren in India my be opened, and those who are able to do so will give us as much literary help as they can.

Praying that the peace of God and His richest blessing may be with you here and hereafter,

I am...

Your obedient brother,

Alexander Russell Webb

APPENDIX F.

WELCOME ADDRESS FROM THE ANJUMAN-TAIDE-MUSULMANAN-E-JADID OF BOMBAY

To Alexander Russel Webb, Esq.[*]

Sir,

We, the members of the Anjumane-Taide-Musulmanan-e-Jadid of Bombay, beg to welcome you to our city as our brother in Islam, and congratulate you as the first American gentleman to embrace the Musulman faith. It is indeed a high honour for you to be a pioneer in the propagation of Islam in the United States of America, and we beg to assure you that we fully sympathise with the object you have in view of establishing a Mahomedan Mission at New York or Chicago. The fact that you have voluntarily sustained a serious loss by resigning your honourable and lucrative post of Consul-General at Manilla conclusively proves the deep interest you take for establishing a Mission on the American Continent, and we are fully satisfied that the said work in your hands will bear good fruits. Ever since you first commenced correspondence with Mr. Budrudin Abdulla Kur of our city, your name has been before the Musulman public of this city, and as an humble token of the deep regard we entertain towards you and the high estimation in which we hold you, we beg on this occasion to present to you this address, conveying our

---

[*] Reprinted from: "Dinner and Address to Mr. A.R. Webb," *Times of India*, November 3, 1892, p. 6.

sincere prayers for your success in the good work you will be shortly undertaking in the United States of America, and we feel sure the Almighty and Merciful Lord of all the worlds will grant you health, strength, and guidance, to enable you to fearlessly discharge the responsible but blessed duties above referred to.

Moulvi Ubaidullah, President
Ghulam Muhammad Munshi, Secretary
Bombay 2nd November 1892

APPENDIX G.

LETTERS TO THE EDITOR OF
THE TIMES OF INDIA

From "A Mahomedan"
November 15, 1892

Sir,

Kindly insert the following few lines in your valuable paper. Now-a-days the ideas of all the Mahomedans of Bombay are directed towards Mr. Webb, a new American Mahomedan. The gentleman wants, as I have heard, only eighty thousand rupees to start a mission in America to propagate the Moslem Faith and to improve the religious state of his brother Americans by raising money especially from the Mahomedans of India. But experienced men know thoroughly well that there are lots of poor and really needy Mahomedans in India, who can hardly earn two annas a day, and thereby they pass their days in half-starvation. Now I consider that it is the real duty of a staunch Mussulman to improve the condition of his co-religionists in India first. Further, I can say that eighty thousand rupees can be safely used in improving the above stated miserable condition of many Mahomedans here. Now I leave the matter to the consideration of my co-religionists, and hope that before doing something for people of a very distant land, they will engage their thoughts in doing something for their brother Mahomedans who are near at hand.

A Mahomedan

From Budruddin Abdulla Kur
November 16, 1892

Sir,

  With reference to a letter appearing in your to-day's issue under the *nom de plume* of "A Mahomedan," you will kindly allow me to lay bare a few facts for the information of the public.

  I have been in correspondence with Mr. Webb for more than two years past, and have been creating interest among the Mussulmans of this city for sending a Mahomedan Mission to the American Continent. So I showed the letters of Mr. Webb, with his permission, to many of my Mussulman friends, and among them to my friend Haji Abdulla Arab. Mr. Haji Abdulla Arab took up the matter so zealously that he went to Manilla at his own expense to confer with Mr. Webb. Mr. Webb having expressed his desire to resign his post of Consul-General at Manilla and devote his whole life to the propagation of Islam in the United States of America, Mr. Haji Abdulla Arab on his part promised to help the American Mission scheme so as to bring it on a sound financial basis. Mr. Abdulla also requested Mr. Webb to visit India, Egypt, Turkey, and Liverpool, on his way to America, so that he might come in friendly contact with the Mussulmans of those countries. This, in short, is the whole story.

  It will thus be seen that Mr. Webb has not come to India to collect and money, but on the contrary, he was requested to come here. Such of the Mahomedans as are willing to show the true light and essence of Islam to the Western world may assist in this noble cause and give their humble quota.

  I think, Sir, that your correspondent should have ascertained facts and made due enquiries on the subject before he rushed into print. I have ample documentary proof to corroborate my statements, and your correspondent may do well to come to me and personally see them instead of writing under a disguise of "A Mahomedan." Whether it be to bring the civilized nations of the West in close fraternity with the Mussulmans or to alleviate the sufferings and the misery of the poor Mussulmans of this country, the field is, indeed, a boundless one for the charitably disposed rich Mahomedans. But will our people wake up and do something?

Yours, &c.

Badrudin Abdulla Kur

From "A Mahomedan"
November 17, 1892

Sir,

With reference to the letter from Mr. Badrudin Abdula Kur, appearing in your issue of to-day, I can, without hesitation, say that the above mentioned gentleman has not understood my letter.

I have not said that Mr. Webb has come to collect money. But Mr. Kur and other enthusiastic and enterprising Mahomedans like himself, who are promoting a movement among Mahomedans to send a Mission to America under the leadership of Mr. Webb, were furnished by Mr. Webb with a memo, showing that eighty thousand rupees will be required for the purpose.

These Indian rupees either Mr. Kur or his colleagues want to use in America to improve the religious state of Christians there. But there are lots of Mahomedans in India, who use the name of the Prophet and God in swearing only, while they know nothing of their religion. Such people can never be considered to be Mahomedans; therefore, their religious state should first be improved, and the eighty thousand rupees would be better spent in India in improving the religious state of the ignorant Indian Mussalmans.

Further, I can say that Mahomedans should look first to the interest of those who are at hand. According to a popular maxim, "Charity begins at home," and Mr. Kur himself affirms that it is a boundless field for charitably disposed rich Mahomedans undoubtedly to give large sums in charities. Therefore we Indian-born Mahomedans should first improve by every means the religious and social state of the really needy and poor Mussalmans, and after satisfying their wants in both ways the surplus money should be used elsewhere; but not till then.

I think that the work of the American Mission will be to propagate Mahomedanism among American Christians, but when there are lots of Mahomedans, who are, in fact, no Mahomedans, we Mahomedans of India should first start a mission here to improve the religious state of the abovementioned people, and then work in a far distant country. Many Mahomedans of India really follow no religion, while enthusiasts like Mr. Kur stretch their eyes to a far distant country although they ignore what happens near their feet.

A Mahomedan

From Budruddin Abdulla Kur
November 22, 1892

Sir,

I am the last person to think that your paper is the proper instrument for the discussion of the preferential claims and merits of any particular sectarian or religious propaganda over any number of schemes that may be put forward against it by individuals screening themselves under a veil of anonymity. However, as you have opened your columns for the discussion of this subject, you will allow me to state the view I take of this matter.

There is a second letter in your issue of to-day from your correspondent "A Mahomedan," and therein he denies having said that Mr. Webb had come to collect money. Now, Sir, as I have no desire to indulge in logomachy and wage a mere war of words, I shall quote a sentence from the first letter. Writing about Mr. Webb, he says, "The gentleman wants, as I have heard, only eighty thousand rupees to start a Mission in America to propagate the Moslem Faith and to improve the religious state of his brother Americans by raising money, especially from Mahomedans of India." These

289

words speak for themselves. Indeed, it is no easy task to coin euphemisms that will soften or conceal the pure meaning conveyed by simple words. It was at the request of my friend, Mr. Haji Abdulla Arab, that Mr. Webb drafted a scheme which is now being so well supported by the Indian Mussulmans.

Further on, your correspondent says that I, or my colleagues, want to use the money raised here to improve the religious state of the American Christians. Now this is a misguiding assertion. I have no supervising hand in the financial management of the scheme, except so far that I think Mr. Webb to be worthy of our confidence, and that those who are of similar opinion need lose no time in extending both literary and pecuniary help towards this movement. The question of the management of the funds is one to be disposed of by the subscribers, and Mr. Webb among themselves, and outsiders need not trouble themselves about it at present.

Your correspondent recommends a number of schemes to the sympathy of the Indian Mussulmans and the speedy rejection of the single one framed by Mr. Webb and supported by enthusiasts like myself. In his first letter he pleaded the cause of poverty-stricken Mahomedans, but in his second letter he goes so far as to ostracise many Indian Mussulmans from the very pale of Islam. He would, therefore, establish a Mission here to bring them back from a wrong path; perhaps, apostasy, and otherwise improve their religious condition. "A Mahomedan" is welcome to do something practical to improve the religious condition of the "many Mahomedans of India who really follow no religion," and to ameliorate the condition of the poor Mahomedans. So he comes out in an anonymous garb to write in the newspapers and begins to work for the noble schemes he has so much at heart, but alas! he works on paper only. May I request him not to disturb other workers in a different channel?

I should like to say a few words about my enthusiasm for the promotion of this American scheme. In order to emphasize my reason in support of this scheme, I shall place before your readers the results of the Liverpool Mission.

About two years ago Mr. Quilliam, of Liverpool, appealed to the Indian Mussulmans to assist him to place the

290

Liverpool Moslem Institute on a substantial basis. As I understand, within this period not more than forty thousand rupees have been remitted to England from different Mahomedan centres in India, with the result that we have now more than two hundred converts in England, and branch societies have been opened in Manchester and London. But, Sir, I do not at present set so high a value upon the number of converts as upon the reflex educational influence made upon the Indian Mussulmans by the establishment of the Liverpool Mission. We have already there a Mosque and a High School, where Mahomedan boys are being sent from India to receive a liberal education, and where religious instruction is not neglected.

We see this day in India a rare and a pitiable phenomenon of a powerful nation of sixty millions of Mussulmans absorbed in apathy and lethargy, and steeped in ignorance, in so far at least as secular education is concerned. Now, who can rouse such a vast number of our Mahomedan brethren to a sense of their duty in educational matters in a surer and a better way than the civilized English and American Mussulmans? For these people can easily make us realise our degraded and fallen condition, and bring about a revival of learning in our community. I have not the least doubt that the noble instincts of the English race, when permeated with the sublime doctrines of Islam, will influence and appeal to the imagination of the Indian Mussulmans to cast aside their languor and the sleep of centuries and take a new place in the roll-call of nations.

I have no desire to magnify the results of either the English or the American scheme. But, looking at the question in a purely educational aspect, I feel that whatever amount of money may be spent on them, will be received back a million-fold at no distant date.

The problem of Indian poverty is one that has baffled solution even at the hands of Government. It is a herculean task, and no one can dream of combating the poverty of our people by a paltry sum of eighty thousand rupees, while the question of the social evils existing among us is one that can be satisfactorily settled by the benign influence of education alone. In short, the one panacea for all the difficulties and the disabilities we labour under is the revival of learning among our people. Poverty is the result of the ignorant condition of our people, as knowledge will be the father of our wealth. Let us strike at the root of ignorance, and our people in course of time will earn, economise, save, and prosper. The pages of history show us many examples of nations that have dwindled down, decayed and perished on account of their ignorant condition, but a nation will not be totally ship-wrecked or effaced from the pages of history, simply because it is poor.

For the educational development and progress of the Indian Mussulmans in their present helpless condition, my eyes are hopefully riveted in India upon the grace of the British Government and in the West upon the civilized American and English converts, whose help and fresh zeal will be of inestimable value in the noble cause of regeneration and revival.

Yours, &c.

Badrudin Abdulla Kur

GLOSSARY

&c. Et cetera.

A's. Annas.

Anjuman. Society or association.

anna. Currency, 1/16 of a rupee.

asotea. Roofless veranda.

Asr. Obligatory afternoon prayer for Muslims.

B.I.S.S. British India Steam Ship (ship associated with the British India Steam Navigation Co.)

brougham. One-horse carriage for two or four passengers.

betel. Asian pepper whose leaves are chewed as a stimulant.

bucksheesh. Handout.

bullock-cart. Large cart pulled by an ox.

bunder. Boat landing.

C.I.E. Companion of the Order of the Indian Empire.

C.S.I. Companion of the Order of the Star of India.

carromata. Philippine two-wheeled cart pulled by a horse.

catafalque. Funerary platform for a coffin or corpse.

cheroot. Indian cigar.

chunam. Indian plaster made of lime and sand.

companionway (companion-way). Stairway area of a ship.

copra. Dried coconut meat.

Dak. Accommodation located on an Indian letter carrier route.

dervish. Ascetic.

desayuno. Breakfast in Spanish.

draft. Order for payment of money.

Escolta. Main area of commerce in Manila.

Esplanade. Public promenade.

F.T.S. Fellow, Theosophical Society.

Fajr (Fujr). Obligatory morning prayer for Muslims.

fakir. Indigent Muslim beggar.

flageolet. Small flute.

G.C.M.G. Knight Grand Cross of St. Michael & St. George.

gharry (garry). Horse drawn carriage for hire.

ghat. Steps on a riverbank leading to the water.

gymkhana. Place for equestrian races and games.

H.H. His Highness.

H.P.B. Helena Petrovna Blavatsky.

H.S. Hyderabad State.

hackman (hack-man). Carriage driver.

Hajee (haji, hadjee). A title for a Muslim who has completed the pilgrimage to Mecca.

half-caste. Mixed race, usually Indian and European.

Hindoo (Hindu). Adherent of the predominate religion in India.

Hindustannee—SEE: Urdu.

hookah. Water-based smoking pipe.

howdah. Seat for riding elephant.

hubble-bubble—SEE: hookah.

Isha. Obligatory night prayer for Muslims.

isinglass. Gelatinous substance.

jinriksha. Light two-wheeled cart drawn by a man.

joss sticks. Incense burned as prayer to ancestors.

Jung (or Jang). Honorific title.

K.C.B. Knight Commander of the Bath.

Kafir. Non-believer.

Kling. Generic name for Indians in Malay countries sometimes used pejoratively.

Koran. Muslim holy book.

landau. A four-wheeled carriage with top that can be opened or closed.

lingham. Phallus.

Macadamized (Macadamed). Named for John MacAdam, refers to a nineteenth-century layered road building method which produced smoother surfaces.

Magrib (Magrb). Obligatory evening prayer for Muslims.

mahout. Elephant driver.

Maidan. Open park or market.

Malay (Maylay). People of Malaysia and some nearby countries who chiefly speak the Malay language.

mayordormo (mayor-dormo). Steward.

Mohammedan. Errant term applied to Muslims, generally used by non-Muslims.

Mohrum. Islamic New Year.

Moslem. Variant of Muslim.

mosque. Muslim house of worship.

Moulvi. Used as a form of address to indicate a learned man or teacher.

Mowlood. Celebration of the birth of the Prophet or a saint.

Muezzin. Performs the Muslim call to prayer.

Muslim. Follower of the Islamic faith.

Mussalman. Variant of Muslim.

296

N.G. No good.

Nautch. Dance performed by professional dancing girls.

Nawab. Title for a Muslim of status.

Nizam. Hereditary title of the ruler of Hyderabad.

P.O. Post Office.

pan-supari. Consists of both the betel leaf and betel nuts, which are chewed.

Parsee. Descendents of Persians living in India who practice Zoroastrianism.

pice. Currency, ¼ of an anna.

pleader. A legal advocate.

precipitate. Spirit taking a human form.

purdah. Practice of screening women from men with a curtain or other barrier.

R's. Rupees.

R.E. Royal Engineer.

R.R. Railroad.

Rajput. Militaristic Hindu caste.

Ramzan. Muslim holy month of fasting (usually spelled Ramadan).

rupee. Silver coin, standard currency of British India.

S.E. Southeast.

S.S. Steamship.

saguan. Walkway to a courtyard.

Sambayan. Filipino term for festival.

sampan. Small slender boat.

Sanscrit. Classical and sacred language of India.

Sepoy. Indian soldier.

Shia. Minority Muslim sect.

Singalese. Variant of Sinhalese, natives of Sri Lanka (Ceylon).

subscription. Contribution.

Sufi. Muslim mystic.

Sultan. Ruler of a Muslim country.

Syce—SEE: hackman.

T.S. Theosophical Society.

Tamil. People native to southeastern India and parts of Sri Lanka.

tank. Reservoir.

tariff. List of rates.

Theosophy. System that focuses on the mystical features of philosophies and religions.

tiffin. Midday meal.

toddy. Coconut palm sap used to make beverages.

tom-toms. Asian drum.

Urdu. Language spoken mostly by the Muslims of northern India.

Victoria. Four-wheeled carriage with seating for two passengers.

zenana. Part of a house reserved for women.

Zohar. Obligatory noon prayer for Muslims.

BIBLIOGRAPHY

NEWSPAPERS

Arizona Republican
Brooklyn Eagle
Chicago Tribune
Constitution (Atlanta, GA)
The Crescent (Liverpool, Eng.)
Daily Gazette and Bulletin (Williamsport, PA)
Evening Democrat (Warren, PA)
Evening Record (Bergen County, NJ)
The Moslem World (New York, NY)
The Moslem World and Voice of Islam (Ulster Park, NY)
New York Daily Tribune
New York Times
The Path (London, Eng.)
St. Louis Republic
Salem Daily News (Salem, OH)
Republican (Unionville, MO)
Rutherford Republican (Rutherford, NJ)
Times of India (Bombay, India)
Trenton Times
Washington Post

BOOKS AND ARCHIVAL MATERIALS

Abbasi, Muhammad Y., comp. *Annals of the Central National Mahommedan Association, 1878-1888*. Islamabad: National Institute of Historical and Cultural Research, 1992.

Abdul Latif. *Autobiography and Other Writings of Nawab Abdul Latif Khan Bahadur*. Edited by Md. Mohar Ali. Chittagong: Mehrub Publications, 1968.

Adams, Oscar F. *A Dictionary of American Authors*. New York: Houghton, Mifflin and Co., 1897.

Ahmad, Basharat. *Mujaddid Azam*. Trans. Hamid Rahman. Forthcoming.

Ahmad, Ghulam. *Shahna-i-Haq*. Qadian: [s.n.], 1888.

Ahmad, Ghulam. *The Teachings of Islam*. London: Luzac, 1910.

"Alex R. Webb Statement of Age, Education, Employments, etc." 1887, Letters of Application and Recommendation, 1885-1893, Record Group 59, 250/47/23/4 Box 131, National Archives and Records Administration.

American National Biography. New York: Oxford University Press, 1999.

Aubertin, John J. *Wanderings & Wonderings*. London: K. Paul, Trench, Trübner & Co., 1892.

Austin, Allan D. *African Muslims in Antebellum America: Transatlantic Stories and Spiritual Struggles*. New York: Routledge, 1997.

Aziz, K.K., ed. *Public Life in Muslim India, 1850-1947*. Lahore: Vanguard, 1992.

Bennett, D.R.M. *Anthony Comstock: His Career of Cruelty and Crime*. New York: Da Capo Press, 1971.

Berger, Arthur S. and Joyce Berger. *The Encyclopedia of Parapsychology and Psychical Research*. New York: Paragon House, 1991.

Blavatsky, H.P. *The Theosophical Glossary*. Los Angeles: Theosophy Company, 1930.

Blunt, Wilfrid S. *India Under Ripon*. London: T.F. Unwin, 1909.

Buckland, C.E. *Dictionary of Indian Biography*. London: S. Sonnenschein & Co., 1906.

Cady, John F. *A History of Modern Burma*. Ithaca: Cornell University Press, 1958.

Campbell, A. Claude. *Glimpses of the Nizam's Dominions*. Bombay: C. B. Burrows, 1898.

Chew, Ernest C.T. and Edwin Lee, eds. *A History of Singapore*. Singapore: Oxford University Press, 1991.

The Chronology of Modern Hyderabad, 1720-1890. Hyderabad: Central Records Office, 1954.

Commercial Directory of Manila. Manila: [n.p.], 1901.

Conger, Charles G.B. *A Record of the Births, Marriages and Death of the Descendants of John Conger of Woodbridge, N.J.* Chicago: S. Smith, 1903.

Cranston, Sylvia. *HPB: The Extraordinary Life and Influence of Helena Blavatsky, Founder of the Modern Theosophical Movement*. New York: Putnam, 1993.

Curtis, Edward E. *Islam in Black America: Identity, Liberation, and Difference in African-American Islamic Thought*. Albany: SUNY Press, 2002.

The Cyclopedia of India: Biographical, Historical, Administrative, Commercial. Calcutta: Cyclopedia Pub. Co., 1907-08.

Dannin, Robert. *Black Pilgrimage to Islam*. New York: Oxford University Press, 2002.

Davis, Moshe, ed. *With Eyes Toward Zion*. Volume II. New York: Praeger, 1983.

Despatches from U.S. Consuls in Manila, Philippine Islands, 1817-1899; (National Archives Microfilm Publication T43, Reels 8-10), Records of the Department of State, Record Group 59.

Diouf, Sylviane A. *Servants of Allah: African Muslims Enslaved in the America*. New York: New York University Press, 1998.

Drake, P.J. "Southeast Asian Monies and the Problem of a Common Measure, with Particular Reference to the Nineteenth Century." *Australian Economic History Review* 31: 90-96.

Draper, John William. *History of the Intellectual Development of Europe*. New York: Harper & Bros., 1876.

Drury, Nevill. *Dictionary of Mysticism and Esoteric Traditions*. Santa Barbara: ABC-CLIO, 1992.

Edwards Annual Directory of Chicago. Chicago: R. Edwards, 1868-70 and 1873. Microfilm.

Faust, Karl Irving. *Campaigning in the Philippines*. San Francisco: Kicks-Judd Co., 1899.

Field, Eugene. *The Complete Tribune Primer*. Minneapolis: Dillon Press, 1967.

Gomez, Michael A. *Black Crescent: The Experience and Legacy of African Muslims in the Americas*. New York: Cambridge University Press, 2005.

Gould's St. Louis Directory. St. Louis: Polk-Gould Directory Co., 1877-82 and 1886. Microfilm.

Grebsonal, L. "The Mohammedan Propagandist." *Frank Leslie's Illustrated Weekly* (March 30, 1893): 204-05.

Green, Nile. "Migrant Sufis and Sacred Space in South Asian Islam." *Contemporary South Asia* 12: 493-509.

Gullick, J.M. *Rulers and Residents: Influence and Power in the Malay States, 1870-1920*. Singapore: Oxford University Press, 1992.

Haberly, Loyd. *Newspapers and Newspaper Men of Rutherford*. [n.p.]: Rutherford Committee, 1964.

Haddad, Yvonne Y. and John L. Esposito, eds. *Muslims on the Americanization Path?* New York: Oxford University Press, 2000.

Haddad, Yvonne Y. and Jane I. Smith, eds. *Muslim Communities in North America*. Albany: SUNY Press, 1994.

Haddad, Yvonne Y., ed. *The Muslims of America*, new edition. Oxford: Oxford University Press, 1993.

Hardy, P. *The Muslims of British India*. Cambridge: Cambridge University Press, 1972.

Havell, E.B. *A Handbook to Agra and the Taj, Sikandra, Fatehpur-Sikri and the Neighbourhood*. New York: Longmans, Green, and Co., 1904.

History of Adair, Sullivan, Putnam and Schuyler Counties, Missouri. Chicago: Goodspeed Pub. Co., 1888.

Howard, J.J. and F.A. Crisp, eds. *Visitation of England and Wales*. London: [Privately printed], 1893-1921.

Hyde, William and Howard L. Conrad, eds. *Encyclopedia of the History of St. Louis: A Compendium of History and Biography for Ready Reference*. St. Louis: Southern History Co., 1899.

Hyderabad in 1890 and 1891: Comprising All the Letters on Hyderabad Affairs Written to the Madras "Hindu" By Its Hyderabad Correspondent During 1890 and 1891. Bangalore: Caxton Press, 1892.

Imadi, Saidul Haq. *Nawab Imad-ul-Mulk: Social and Cultural Activities of Nawab Imad-ul-Mulk Syed Husain Bilgrami in Hyderabad.* Hyderabad: State Archives, 1978.

The Imperial Gazetteer of India. New Ed. Oxford: Clarendon Press, 1908.

Ives, Nadirah F. "The Story of the First American Convert to Islam." *The Light* (April 8, 1994): 6-7and (April 16, 1944): 5-8.

Jain, Naresh Kumar. *Muslims in India: A Biographical Dictionary.* New Delhi: Manohar, 1979-83.

Jain, Satish Kumar. *Progressive Jains of India.* New Delhi: Shraman Sahitya Sansthan, 1975.

Johnson, K. Paul. *The Masters Revealed: Madame Blavatsky and the Myth of the Great White Lodge.* Albany: SUNY Press, 1994.

Kelsoe, William A. *St. Louis Reference Record: A Newspaper Man's Motion-picture of the City When We Got Our First Bridge, and of Many Later Happenings of Local Note.* St. Louis: Von Hoffmann, 1927.

Khalidi, Omar. *An Indian Passage to Europe.* Forthcoming.

Khan, Agha Mirza Beg. *My Life: Being the Autobiography of Nawab Server-ul-Mulk Bahadur.* Trans, Nawab Jiwan Yar Jung Bahadur. London: Stockwell, 1932.

Kőszegi, Michael A. and J. Gordon Melton. *Islam in North America: A Sourcebook.* New York: Garland, 1992.

Lane-Poole, Stanley. *The Story of the Moors in Spain.* New York: G.P. Putnam's Sons, 1890.

Leonard, Laren I. *Muslims in the United States: The State of Research.* New York: Russell Sage Foundation, 2003.

Lethbridge, Roper. *The Golden Book of India: A Genealogical and Biographical Dictionary of the Ruling Princes, Chiefs, Nobles, and Other Personages, Titled or Decorated, of the Indian Empire, with an Appendix for Ceylon.* London: S. Low, Marston & Co., 1900.

"Letters of Application and Recommendation, 1885-1893." Record Group 59, 250/47/23/4 Box 131, National Archives and Records Administration.

"List of American Citizens,"*Despatches from U.S. Consuls in Manila, Philippine Islands, 1817-1899*; January 1, 1891,

(National Archives Microfilm Publication T43, Reel 10), Records of the Department of State, Record Group 59.

Lo, Mbaye. *Muslims in America: Race, Politics, and Community Building*. Beltsville, MD.: Amana, 2004.

Maitra, Jayanti. *Muslim Politics in Bengal, 1855-1906: Collaboration and Confrontation*. Calcutta: K.P. Bagchi, 1984.

Majumdar, R.C. *The History and Culture of the Indian People*. Volume 6. London: Allen & Unwin, 1951-.

Martin, Richard C., ed. *Encyclopedia of Islam and the Muslim World*. New York: Macmillan Reference, 2004.

Master Hands in the Affairs of the Pacific Coast. San Francisco: Western Historical, 1892.

McCloud, Aminah B. *African American Islam*. New York: Routledge, 1995.

Meade, Marion. *Madame Blavatsky: The Woman Behind the Myth*. New York: Putnam, 1980.

Mehra, Parshotam. *A Dictionary of Modern Indian History, 1707-1947*. Delhi: Oxford University Press, 1987.

Miller, Stephen B. *Historical Sketches of Hudson, Embracing the Settlement of the City, City Government, Business Enterprises, Churches, Press, Schools, Libraries, &c*. Hudson: Bryan & Webb, 1862.

Mudiraj, K. Krishnaswamy. *Pictorial Hyderabad*. Hyderabad: Chandrakanth Press, 1929.

Muhammad, Amir N. *Muslims in America: Seven Centuries of History, 1312-2000*. Beltsville, Md.: Amana, 2001.

Müller, F. Max. *Natural Religion: The Gifford Lectures Delivered Before the University of Glasgow in 1888*. New York: AMS Press, 1975.

Mullin, Donald, comp. *Victorian Actors and Actresses in Review: A Dictionary of Contemporary Views of Representative British and American Actors and Actresses, 1837-1901*. Westport, Conn.: Greenwood Press, 1983

Mulliner, K. and Lian The-Mulliner. *Historical Dictionary of Singapore*. Metuchen, N.J.: Scarecrow Press, 1991.

Nyang, Sulayman S. *Islam in the United States of America*. Chicago: ABC International, 1999.

Paine, S.C.M. *The Sino-Japanese War of 1894-1895: Perceptions, Power and Primacy*. Cambridge: Cambridge University Press, 2003.

Platts, John T. *A Dictionary of Urdu, Classical Hindi, and English*. London: W. H. Allen & Co., 1884.

Raj, Sheela. *Mediaevalism to Modernism: Socio-economic and Cultural History of Hyderabad*. Bombay: Popular Prakashan, 1987.

Representative Men of the Bombay Presidency. Bombay: C.B. Burrows, 1900.

Richards, John F. *Mughal Administration in Golconda*. Oxford: Clarendon Press, 1975.

----. *The Mughal Empire*. Cambridge: Cambridge University Press, 1993.

Rizal, José. *Noli Me Tangere*. Translated by Ma. Soledad Lacson-Locsin. Honolulu: University of Hawai'i Press, 1997.

Saeed, Ahmad. *Muslim India (1857-1947): A Biographical Dictionary*. Lahore: Institute of Pakistan Historical Research, 1997.

Scharf, J. Thomas. *History of Saint Louis City and County, from the Earliest Periods to the Present Day*. 2 vols. Philadelphia: Louis H. Everts & Co., 1883.

Shahid, Maulana Dost Mohammed. "Review of Religions: A 100 Year History of the Magazine." *Review of Religions* 97 (November 2002): 7-59.

Singh, Nagendra K. *Encyclopaedia of Muslim Biography: India, Pakistan, Bangladesh*. New Delhi: A.P.H. Publishing, 2001.

Singleton, Brent. "The Ummah Slowly Bled: A Select Bibliography of Enslaved African Muslims in the Americas and the Caribbean." *Journal of Muslim Minority Affairs* 22 (2002): 401-412.

Slout, William L. *Chilly Billy: The Evolution of a Circus Millionaire*. San Bernardino, CA: [William L. Slout], 2002.

Smith, Jane I. *Islam in America*. New York: Columbia University Press, 1999.

Song, Ong Siang. *One Hundred Years' History of the Chinese in Singapore*. Singapore: Oxford University Press, 1984.

South-East Asian Biographical Archive. München: K.G. Saur, 2002. Microfiche.

Spencer, Herbert. *First Principles*. New York: Appleton, 1898.

Stevens, Walter B. *100 Years of the St. Louis Republic*. St. Louis: Republic, 1908.

Times of India Calendar and Directory. Bombay: The Times of India Office, 1894.

Tregonning, K.G. *Home Port Singapore*. Singapore: Oxford University Press, 1967.

Tunison, Emory H. "Mohammed Alexander Russell Webb, First American Muslim." *The Arab World* 1 (1945): 13-18.

Turnbull, C.M. A *History of Singapore: 1819-1988*. Singapore: Oxford University Press, 1989.

Turner, Richard B. *Islam in the African-American Experience*. Bloomington: Indiana University Press, 2003.

Tyabji, Husain B. *Badruddin Tyabji: A Biography*. Bombay: Thacker & Co., 1952.

Upton, Charles Williams. *The History of the Hudson (New York) Daily Star, 1847-1940: The Biography of a Small City Newspaper*, Ph.D. diss., Harvard Dissertation, 1942.

Volwiler, Albert T., ed. *The Correspondence Between Benjamin Harrison and James G. Blaine, 1882-1893*. Philadelphia: American Philosophical Society, 1940.

Wallace, W. Stewart. *A Dictionary of North American Authors Deceased Before 1950*. Detroit: Gale, 1968.

Webb, Alexander Russell. "Two Remarkable Phenomena." *New Californian* 1 (January 1892): 248-51.

----. "Islam and Theosophy." *Lucifer* 10 (July 15, 1892): 421-25.

----. *The Three Lectures*. Madras: Lawrence Asylum Press, 1892.

----. *Islam in America: A Brief Statement of Mohammedanism and an Outline of the American Islamic Propaganda*. New York: Oriental Publishing Co., 1893.

----. *Lectures on Islam: Delivered at Different Places in India*. Lahore: Mohammadan Tract and Book Depot, 1893.

---. "Preaching Islamism in America." *Providence Journal* 14 (November 1893): 463-70.

Who Was Who in America. Chicago: Marquis Co., 1942.

Who's Who in India: Containing Lives and Portraits of Ruling Chiefs, Notables, Titled Personages, and Other Eminent Indians. Lucknow: Newul Kishore Press, 1911.

Winters, Christopher, ed. *International Dictionary of Anthropologists*. New York: Garland, 1991.

Worcester, Dean C. *The Philippine Islands and Their People: A Record of Personal Observation and Experience, with a Short Summary of the More Important Facts in the History of the Archipelago*. New York: Macmillan, 1899.

---. *The Philippines Past and Present*. New York: Macmillan, 1914.

Worswick, Clark, ed. *Princely India: Photographs by Raja Deen Dayal, 1884-1910*. New York: Knopf, 1980.

INDEX

Muslims in Wadi—194-195, 205
Mustard, Mr. & Mrs.—50, 54-55
Nabakoff, Emin L. (Leo)—34-35
Nation of Islam—45-47
Natunas Islands—60
Nepaulese Temple—132
Nizam (of Hyderabad)—165, 168-169, 171, 175-176, 184, 187, 189-193
Nizam Club—162, 185, 187-188, 193
Nizam's Dominions—24, 160-161, 205
O' Bryant, William—9-10
Olcutt, Col. Henry S.—15, 24, 119, 133, 197-199, 201
Omar, Khalif—224
Order of Medjidie—42
Oriental Literary Bureau—33
Oriental Publishing Company—28
Oudh, King of—111, 120
Oudh, Prince of—116, 127
Padgham, Reverend Elizabeth—44
Pan supari—166
Paraceleus—198
Paragua—57-58
Parsees—104; in Bombay 140, 144-145, 148, 150
Patna—130
Patna Institute Gazette—130
Peabody, H. Fatima—38n
Pegging—177
Penang—84-85; Chinese in 85
Perambur—197, 204
Peshwa—157
Petrarch—250
Petty-Fitzmaurice, Viceroy Henry Charles Keith—113
Philippines—18-22, 50-55, 227-228, 234
Philosophers—211, 231, 236, 239
Plowden, Sir Trevor—175n
Polygamy—32, 220, 222-223, 245, 272-273
Poona—153, 155-160; Muslims in 156, 158
Pope John XXII—250
Portuguese—147-148
Prayer—257-262
Predetermination—213
Presbyterians—5, 209, 270
Prime Minister of Hyderabad—SEE: Asman Jah, Sir
Prophets—212, 217-218, 252, 256
Prostitution—74-75, 79, 151, 245
Public Garden (Hyderabad)—170, 175-177, 185, 187, 189-190, 195
Public Museum (Calcutta)—122
Publishing—28-30, 36, 38-39
Punch—166

316

Purdah—153, 168, 245
Quilliam, Sheik Abdullah W.H.—294
Qur'an—16, 32, 93, 116, 150-151, 162, 166-167, 187, 212, 219-220, 226, 242, 244, 251-252, 254-255, 263-265, 270, 274, 282
Raichur—205
Rangoon—89-106; Chinese in 101-102; Indians in 96; Muslims in 40, 89-91, 93-94, 96-98, 100, 103-104, 106
Raffles Hotel—62-63
Raffles Library and Museum—73
Rail Travel—129, 131, 134-135, 155, 160-161, 193-197, 204-206
Rangoon—23
Rawson, Albert Leighton—28
Raza, Hafiz Ahmed—170
Regatta—169-170
Reis & Rayyet—124
Religion, definitions of—232-234
Republican (St. Louis, Mo.)—14, 97n, 277
Republican (Unionville, Mo.)—9-11, 277
Ridley, Henry Nicholas—78-79
Ripon, Lord—113
Robson and Crane—14
Royal Asiatic Library—139
Rutherford, N.J.—42, 44
Rutherford News—42
Sachin—142
Saint, Muslim—181
Sale, George—246
Salvation Army—220, 270
Sambayan—75
Séance—181-182
Secunderabad—166-167, 172, 181, 184-185, 189
Server Jung, Nawab—168, 191
Shah Sahib—SEE: Abdur Rahim, Syed Shah
Shahabad—205
Shahabudin, Kazi—158-160
Sharful Hak, Mohammed—166-167
Shiahs—128, 140
Ship travel—55-61, 82-84, 87-89, 106-111
Sickels, David B.—19
Simpson, George—50-52
Singapore—19-20, 22-23, 61-82; Buddhists in 67, 69; Chinese in 63-64, 71-72, 74-76, 79-81; Hindus in 68-71, 81; Klings in 81; Malays in 78; Muslims in 63, 69-71, 81; Tamils in 81
Singapore Cricket Club—64-65
Smith, Bosworth—129
Snow, Hamid—35
Soliman, Hajee Yussuf—150-151
Soorunagar—174-175

early politics 10-11, 18, 277; early study of eastern philosophies 14-16, 210, 277, 279; early study of Islam and conversion 16, 19-20, 209-211, 220, 231, 272-273, 279-283, 285-287; jewelry trade 6-8, 13, 276; mission to America 21, 26, 30, 33, 93, 96, 101, 138-139, 175, 244-247, 273-274, 285-295; newspaper business 9-14; press reaction to mission 25-26; Theosophical beliefs 17; travels in Agra 206-208; travels in Benares 131-134; travels in Bombay 23-25, 195-196; travels in Calcutta 111-129; travels in Hyderabad 24, 160-194; travels in India 23-25, 109-208; travels in Madras 197-204; travels in Nizam's Dominions 24, 160-161, 205; travels in Patna 130; travels in Penang 84-86; travels in Poona 155-160; travels in Rangoon 23, 89-106; travels in Singapore 22-23, 61-82; travels in Wadi 160, 193-195, 205; views on British and British colonialism 65-66, 96, 104, 106-107, 111, 115-117, 120, 135, 142, 151, 160-161, 166, 169, 175, 184, 191, 205; views on Indians 143, 146, 152, 161, 164, 166, 168-169, 173, 175, 191; views on Christian missionaries 117, 154, 172, 246-247, 251, 284; views on Christianity 209-211, 218, 220-221, 223-225, 227, 249-253, 270-274, 284; views on class 41; views on race 41, 98, 193, 202; views on religious fanaticism 128, 132, 150, 168, 172-173, 179, 181, 207, 247, 250-251; views on science 210-211, 219, 235-240, 242; views on Spanish and Spanish colonialism 22, 56-60, 228; views on Western civilization 220-221, 230, 251, 260

Webb, Anna May—5
Webb, Bessie (Hotchkiss)—12, 51, 136, 145
Webb, Caroline Elizabeth Lefferts (mother)—4, 11, 149
Webb, Carrie Lefferts (sister)—5
Webb, Clarence Herbert (brother)—8, 11
Webb, E. (Edward) Cook (brother)—4, 7
Webb, Ella G. (Hotchkiss) (wife)—20, 36, 44-45, 50-51, 54, 118, 122-123, 136-137, 140, 145, 149, 169, 200, 272, 286
Webb Family—4
Webb, Herbert Nelson (brother)—4, 7
Webb, Mary (Mamie) Caroline (daughter)—15, 45, 48, 67, 122
Webb, Nala (daughter)—19, 67
Webb, Russell Lorenzo (son)—14, 38
Webb, William Bunker (brother)—4
Wendell and Hyman Jewelers—8, 276
Werner, Edgar S.—39
White, R. Othman—38n, 39n
White House—168
Wildman, Rounceville—63, 66, 78
Worcester. Dean C.—51
World Parliament of Religions—31-32
X, Malcolm—46
Yoosuf, Hajee Cassim—142
Youssuf, Mohamed—122, 126
Yussuf, Hajee Haroun Hajee Jaffer—156-158, 160

www.ingramcontent.com/pod-product-compliance
Lightning Source LLC
Chambersburg PA
CBHW031236090426
42742CB00007B/218